William John Thoms

Human Longevity

Its Facts and its Fictions

William John Thoms

Human Longevity
Its Facts and its Fictions

ISBN/EAN: 9783337365660

Printed in Europe, USA, Canada, Australia, Japan

Cover: Foto ©berggeist007 / pixelio.de

More available books at **www.hansebooks.com**

HUMAN LONGEVITY

ITS FACTS AND ITS FICTIONS

INCLUDING

AN INQUIRY INTO SOME OF

THE MORE REMARKABLE INSTANCES, AND SUGGESTIONS

FOR TESTING REPUTED CASES

ILLUSTRATED BY EXAMPLES

By WILLIAM J. THOMS, F.S.A.

DEPUTY LIBRARIAN, HOUSE OF LORDS

THE NUMBER OF A MAN'S DAYS ARE AT MOST ONE HUNDRED YEARS
Ecclesiasticus xviii. 9.

TO

PROFESSOR OWEN

D.C.L. F.R.S. &c.

MY DEAR PROFESSOR

I VENTURE TO DEDICATE THIS LITTLE BOOK TO
YOU, PARTLY BECAUSE YOUR DISTINGUISHED NAME WILL
SECURE FOR IT AN ATTENTION WHICH IT MIGHT NOT
OTHERWISE OBTAIN; BUT CHIEFLY FROM MY ADMIRATION
OF YOUR GENIUS AND ATTAINMENTS AND MY GRATITUDE
FOR YOUR KINDNESS

YOURS FAITHFULLY AND SINCERELY

WILLIAM J. THOMS

PREFACE.

THE DURATION of HUMAN LIFE has hitherto been
treated almost exclusively by Naturalists and Physiologists—men eminently qualified by their professional attainments to do full justice to the subject, so far as relates
to the scientific conclusions to be deduced from the facts
before them. They have, however, taken as facts what
are really, in the majority of cases, mere assertions, and,
by arguing from false premises, have arrived at very
erroneous and unjustifiable conclusions.

If Old Parr really attained the age of 152, and Henry
Jenkins that of 169, the great German physiologist
Haller might have been justified in arguing, as he has
done, from such data, the possibility of man's life being
extended to two hundred years.

But, if it be shown that there is not a tittle of evidence
to prove that Parr and Jenkins did attain the extraordinary ages with which they have been credited, the theory
based upon their supposed abnormal Longevity necessarily falls to the ground.

The object of the present volume is to examine the important question of Human Longevity from a plain, common-sense point of view. For this purpose the cases of Parr, Jenkins, and some other reputed Centenarians have been treated as any other historical facts should be treated—credited as far as they are susceptible of proof, but not one atom beyond; while, in addition, a number of recent cases have been subjected to similar criticism.

As to the manner in which Human Longevity is discussed in the following pages, I claim no credit for originality. Many years since the late Mr. Dilke (whom I can never mention without an acknowledgment of how much I owe to his friendship and advice) applied his critical spirit and excellent judgment to the exposure of how vast an amount of error and credulity existed on the subject of Centenarianism.

He was followed by another lamented friend, Sir George C. Lewis, who, in 'Notes and Queries' and elsewhere, applied to the investigation of the Duration of Human Life the same intelligent spirit of inquiry which he had brought to bear upon the Mythical History of Rome, and other similar questions.

Had either of these earnest searchers after truth undertaken such a work as the present, much would have been done to correct the popular errors which now prevail upon this subject. I can only hope that some compen-

sation for my inferior ability to treat the question may
be found in the advantage which I enjoy in following in
the steps of such sagacious leaders. ' The dwarf,' says
Coleridge, ' sees further than the giant, when he has the
giant's shoulders to mount on.'

I have this reflection to encourage me ; and I feel
also that in one thing, at least, I do not yield to either
of my eminent predecessors—and that is, in an earnest
desire to ascertain the Truth, the whole Truth, and
nothing but the Truth, upon this very important physiological and social question.

WILLIAM J. THOMS.

40 ST. GEORGE'S SQUARE, S. W.
April 1873.

CONTENTS.

CHAPTER I.

CHAPTER II.

CHAPTER III.

CHAPTER IV.

CHAPTER V.

CHAPTER VI.

xii   *Contents.*

CHAPTER VII.

CHAPTER VIII.

CHAPTER IX.

CHAPTER X.

APPENDICES.

HUMAN· LONGEVITY.

—◦—

CHAPTER I.

WHEN it is considered for how many years Insurance Offices have existed in this country, and, consequently, for how many· years the attention of men of science has been directed to the Rates of Mortality and the average ages to which individuals may be expected to attain, it is certainly somewhat remarkable, that it should be left to writers of the present day to inquire, for the first time, how far the statements of exceptional Longevity, which are so commonly and persistently circulated, are founded in truth.

It would from this seem to be the fact, that the un-hesitating confidence and the frequency with which the public is told of instances of persons living to be a hundred years of age and upwards, so familiarises the mind to the belief that Centenarianism is a matter of every-day occurrence, that the idea of questioning the truth of any such statements never appears to have suggested itself.

After reading, within a short period, of the death of

Ebenezer Baillie, at 103; of Captain MacPherson, at 100; of Betty Evans, at 102; of Mr. John Naylor, at 117; of Mrs. Sarah Edwards, at 104; of Mrs. Margaret Curtis, at 103; of Sarah Pay, at 104; of Sarah Jones, at 108; of Sarah Clarke, at 108; of Matthew Baden, at 106; of Richard Purser, at 112; and Jacob Fournais, at 135; of Jonathan Reeves, still living, at 105; of William Webb, at the same age; of the Parish Officer of Chelsea sending up in a balloon an old woman (Mrs. Hogg) to celebrate her 100th birthday; and of public entertainments to celebrate Susan Stevenson's 100th birthday; and of public breakfasts to Captain Lahrbush, to celebrate his 104th and 105th birthdays,—after reading such announcements as these, and dozens of similar notices, it seems almost an impertinence to doubt the accuracy of any of these statements, though there is probably scarcely one per cent. among these confident announcements which would bear the test of a thorough investigation.

Those only who have undertaken such investigations can form the slightest idea of the difficulties which enquirers into the truth of statements of this kind have to encounter. It is in vain to assure the parties addressed that not the least suspicion is entertained of their good faith and truthfulness; and that all that is suspected is an unintentional error, a confusion between two parties of the same name; the applying to one party of a baptismal certificate, or some similar piece of evidence, which relates to a totally different person. These and similar assurances are, as a rule, pleaded in vain. You have doubted the truth or accuracy of their statements—statements which they '*know to be true*' (but how they know it they would find it difficult to explain) ; and, if it is determined to pursue the investigation further, it becomes necessary to apply for information and assistance to some other quarter.

Yet, in strange contrast with the feeling of indignation so often manifested when doubt is thrown upon any cases of supposed abnormal Longevity, is the confidence and recklessness with which the most startling announcements of such cases are given to the world, without the least preliminary inquiry, and often without a particle of foundation.

Whether the love of the marvellous, which is more or less inherent in all people, blinds their judgment ; or a careless indifference to that accuracy which should be observed in all statements professing to be statements of facts, leads them to disregard the obligation of not asserting as of their own knowledge matters of which they really know nothing, it is certain that many respectable people do not hesitate to declare in the most reckless manner, that they *know* John Nokes to be 105, and Mary Styles 106, when it is perfectly clear that whatever they may believe, they have never taken the trouble to investigate the cases, and really know nothing of the subject.

Let me lay before the reader a few amusing examples of the thoughtlessness—to use the mildest term which is applicable to such conduct—with which statements of this nature are brought before the public. In 'The Times' of January 21, 1867, a writer under the sig-

nature of *Gerontophilos*, in a letter modestly headed
'Longevity—a Challenge,' solicited 'space for (his) statement of the age of the oldest man probably now living
in England,' and then proceeds :—' In the parish of
Leckhampton, adjoining Cheltenham, there is a man of
the peasant class, named Percy. He was born in a
village between Worcester and Malvern. In the spring
of 1861, on his 105th birthday, he dined in my kitchen.
I saw him walking, with the aid of crutches, in Cheltenham, in November last. He was then in his 110th
year, and is, I have reason to believe, now alive. The
proofs of his birth and age were furnished to the
minister of his parish in 1860, and were sent to the
Queen, from whom he received a gratuity of 5*l*.'

Gerontophilos had not even taken the pains to ascertain the correct name of his hero. It was *Purser*, not
Percy. I knew something of the case, and answered
the challenge in the only way I could, by writing to
' The Times,' and asked for the ' proofs of his birth and
age.' From want of space, or for. some other sufficient
reason, my answer was not inserted, and the matter
dropped.

But when the old man died, and was buried at Cheltenham, with this incription upon his coffin : 'RICHARD
PURSER, died 12 October, 1868, aged 112 years'—it
appeared that his certificate of baptism could not be
found, and the evidence of his age rested on the belief
of two ladies—daughters of a former rector of the parish
in which Purser is said to have been born—which they
based on two very inconclusive facts, even if they were

established—and on his statement that he recollected
the illuminations at the Coronation of George III.! I
have a photograph of him, taken when he was supposed
to be 104—but was in all probability, to judge from his
appearance, not more than fourscore and four.

A very similar case has recently occurred. A gentleman, a stranger, knowing the interest I took in such
matters, called my attention to the case which had come
under his personal knowledge :—' In a branch of my
own family there lived and died an old servant who was
100 years in the family. His name was —— —— ; he
came a parish apprentice, and died at the age of 108.'
After some further particulars he referred me to the
clergyman of the parish, who knew the old man well, and
could furnish satisfactory evidence as to his age. I wrote
to the clergyman accordingly, and in due course received
a very polite answer from him, stating that he really did
not know anything of the case, and had searched his
register without finding any such name upon it. I
replied by giving him all the particulars with which my
correspondent had furnished me, when he at once recognised the man, whose name was very different from
that stated by my correspondent,—remembered burying
him, and had no doubt of the correctness of his entry
in the burial registry, which showed that this supposed
Centenarian was but little more than fourscore at the
time of his death.

A very striking instance of this recklessness was
afforded during a comparatively recent correspondence
in ' The Standard.' In that Journal of April 11

1870, appeared a letter from a gentleman, who, after complaining of the 'strange diṣinclination now existing to credit that any person in these latter ages has attained the age of 100 years,' proceeds :—'I beg leave, therefore, to furnish an instance quite beyond any question or doubt. It is that of the Rev. Josiah Disturnell, one of the Grecians at Christ's Hospital, and who left that institution to proceed to Pembroke College, Cambridge, in 1763. He must then have been at least 16 years of age, the earliest period at which the Grecians leave the school. Mr. Disturnell was eventually presented to the Rectory of Wormshill, in Kent, which living is in the gift of the Governors of Christ's Hospital.[1] This clergyman retained the living till his death in 1854, and consequently, he had at least reached the patriarchal age of 107 years.'

'These facts may be easily tested by a reference to the Pembroke College registers and the parish register at Wormshill.'

It seems difficult to believe that a gentleman who could write thus confidently, specifying the dates when Mr. Disturnell proceeded to Pembroke College, and of his death as Rector of Wormshill, could be mistaken in the fact that Mr. Disturnell had 'at least reached the patriarchal age of 107.'

How much he must have been surprised when he took up 'The Standard' of April 13, and read in a letter from Mr. Dunn, a gentleman who had also belonged

---

[1] See Trollope's 'History of the Hospital,' pp. 126.

to Pembroke College, the following statement of the truth :

'My brother-in-law, himself also an old Grecian, and of Pembroke College, Cambridge, was inducted to the Rectory of Wormshill in 1835, on the death of Mr. Disturnell, who consequently died long before 1854 ; and, in addition to this circumstance, I would add that a slab is placed in Wormshill church, giving the age of Mr. Disturnell at the time of his death as either 91 or 93.'

This case furnishes a striking confirmation of the opinion at which we have arrived, after considerable experience in inquiries of this nature, namely, that, as a rule, the less foundation there is for a case of alleged Longevity, the greater is the confidence with which it is brought forward and its truth insisted upon.

And here, perhaps, as conveniently as anywhere, I may be permitted to notice a curious phenomenon with which every inquirer into the question of the average duration of human life is met at his outset. Remembering, as he cannot fail to do—for many of the most eminent members of the medical profession take care to bring it prominently before the public—the scepticism which obtains among them on the subject of our origin and organisation, the inquirer cannot but be struck, as I have been, with the simple child-like faith with which men of the highest eminence in medical science accept without doubt or hesitation statements of the abnormal prolongation of human life,[1] which startle plain matter-

---

[1] These notes on the manner in which medical men, as a rule, have hitherto treated the question of human longevity, formed the subject of a letter which appeared in *The Times* on September 4, 1871.

of-fact men when their attention is called to them. Nay, not only receive these astonishing statements as if they were established and well-authenticated facts, but proceed to use them as premisses from which to draw deductions still more startling.

At this time, when the important questions—What is the average duration of human life? What is the greatest age which any human being has ever attained? —are attracting a good deal of attention, it may not be without benefit to the cause of scientific truth to trace the origin of this curious phase of the medical mind ; more especially since it can scarcely be doubted that the attention of the members of this most intelligent profession being once called to the necessity of ascertaining that the alleged facts are facts, before proceeding to argue from them, a thorough revolution in medical opinion on this important point will be the result.

But it may be doubted whether I am not doing injustice to my medical friends by these remarks. A passage or two from an article on Longevity in the ' Edinburgh Review' for January, 1857, written by one of the most eminent physicians of the day, will justify what I have stated. Sir Henry Holland, whom I may mention by name as the paper is reprinted in his valuable ' Essays on Scientific and other Subjects,' says :—

' At present it is enough to state that we have sufficient proof of the frequent prolongation of life to periods of from 110 to 130 or 140 years—cases which, thus authenticated, we must necessarily take into view when dealing with the question of human longevity.'

In the face of this confident assertion, I feel assured that if its distinguished writer attempts to produce evidence of any human being having attained the age, not of 130 or 140, but of 110 years, that evidence will be found upon examination utterly worthless ; whereas such a fact being directly at variance, not only with all our daily experience, all our life-tables, all the records of our insurance offices, would require to be supported by evidence at once clear, direct, and beyond dispute.

But the writer of the article in question did not recognise this necessity ; for a little further on, after some hesitation as to Henry Jenkins having attained the age of 169 (!), he proceeds :—

' Yet we cannot equally reject the evidence as to the 152 years of Thomas Parr's life, accredited, as it is, by the testimony of Harvey, who examined his body after his death.'

With all due respect to Sir Henry Holland, I contend that Harvey does not bear testimony to Parr's age, but simply records what he was told about it. He was called upon to perform the *post-mortem* examination, and commences his report of such examination with the description of Parr which had been furnished to him.[1]

But, in fact, the Reviewer was in this instance only

---

[1] It was no part of Harvey's duty to ascertain how far the age of the deceased had been accurately stated. Had he done so I feel strongly convinced that he would have struck off many years, probably half a century, from the reputed age of the ' Old, Old, Very Old Man.' Since this was placed in the hands of the printer I have had the satisfaction of seeing this opinion confirmed by no less an authority than Professor Owen. See his article ' On Longevity ' in Fraser's Magazine for February 1872, p. 229.

repeating the rash assertion of the author of the book he was reviewing, M. Flourens, who, in his work ' De la Longévité Humaine,' p. 256, remarks, ' even the timid Haller accepted as certain Parr's age of 152,' and adds, ' *et qui l'est en effet, car il eut pour temoin* Harvey.'

It is difficult to understand on what principle M. Flourens charges Haller with timidity or hesitation in accepting statements of centenarianism, since Haller, who speaks of Jenkins' age of 169 as ' *satis probabiliter*,' has in addition to the cases of Parr and Jenkins, from which he draws the inference that the life of man may be prolonged to 200 years, collected more than a thousand cases of people dying between 100 and 110 ; 60 between 110 and 120 ; 29 between 120 and 130; 15 between 130 and 140; and 6 between 140 and 150.[1]

Had Haller, Flourens, or Sir Henry Holland brought to bear upon these statements the intelligence and acumen with which they would have examined any purely medical or physiological question, they would at once have suspected the inaccuracy of the accounts, tested and rejected them, and so have contributed to the correction of error, instead of to its dissemination. But the habit of receiving, and properly receiving, without hesitation the statements of their scientific brethren

---

[1] In the article to which I have already referred, Professor Owen agrees with my ' estimate of the notes cited by Haller in his "Adversaria" of the thousand cases of longæval individuals between 100 and 150.' They exemplify the patient industry of that voluminous compiler, who gathered all the stray notices of marvellous old people given, as usual, a century or more ago, without any sure or steadfast ground, on such hearsay, self-assertion, and belief, as characterise the cases of Jenkins and Parr.

as to the results of certain experiments, the products of
analyses, the details of operations, and the effects of
remedial agents—matters of fact coming within the
personal knowledge of those who report them, and
whose evidence, therefore, is all that can be required—
leads medical men to receive with the same confidence
statements as to the ages of very old patients, such
statements being, as a rule, unsupported by a particle
of evidence, and founded either on village gossip or on
the confused and fading memories of the old people
themselves.

I am here sorely tempted to say a few words on the
subject of autopsies of supposed Centenarians, even
though by so doing I may lay myself open to the
charge of treating of matters of which I am ignorant.
I will therefore content myself for the present with
cautioning all medical readers of such reports, to be
sure that the writers have taken the sensible advice
of good Mrs. Glass, ' first catch your hare,' and first
ascertained that the subjects of their investigation were
really Centenarians.

How far this caution is from being uncalled for is
very easily proved. In the ' Medical Times' of
March 25, 1871, is an article, ' Autopsy of a Centenarian,' in which the writer describes the post-mortem
of an old fellow, who under the name of Thomas
Geeran, had long imposed upon the good people of
Brighton, as a remarkable instance of abnormal longevity. I had shown in ' The Times,' and elsewhere,
that there was not a · shadow of foundation for this

statement ; but the writer insists on treating the case as that of a Centenarian, although he cautiously describes him at the opening of his paper as being of the reputed age of 105 years and six months. In 'Notes and Queries' of March 2, 1872, will be found a very exhaustive article on this old soldier, in which it is proved almost to demonstration, that his real name was Michael Gearyn, or Gayran, who had enlisted in the 71st on March 3, 1813, and deserted from it on April 10 following, and that his real age was about 83 and not 105. The full particulars of this case, which is a very typical one, will be found in the latter part of the volume.

Another justification of this caution is furnished in the 'Journal of the Anthropological Institute' for April 1872, which contains at pp. 78–87, an article entitled 'The Physical Condition of Centenarians, as derived from Personal Observations in *Nine* [1] Genuine Examples.' The number of Centenarians which the writer has had the opportunity of examining is startling ; and one naturally expects to see the evidence of their great and exceptional age clearly set forth ; but the author is unfortunately reticent upon this important point, and contents himself with speaking of them as 'undoubted examples,' and with asserting 'that of the accuracy of their ages there is not a doubt,' and assuring us of 'his anxiety to satisfy himself upon this

---

[1] They are : 1, Jacob Luning, 103 ; 2, — Eldritch, 104 ; 3, Elizabeth Brown, 101 ; 4, Miss Wallace, 101 ; 5, Ann Hogg, 102 ; 6, Mary Patterson, 101 ; 7, Sarah Stretton, 102 ; 8, Sarah Debenham, 103 ; and 9, Ann Slocomb, 100 ; all then living, except Luning and Brown.

point.' But when it is seen that with regard to one of them, — Eldritch, ' born in the county of Gloucester in July (December 10?) 1767,' that though the man 'can be seen in London,' the writer has not succeeded in learning his christian name, the precise place of his birth—or the precise date. of it—one must feel that on the non-professional part·of his inquiry, the writer has been very easily satisfied.

CHAPTER II.

IT is certainly matter for surprise when we bear in mind how long Man has existed on the face of the globe, that the questions, What is the ordinary duration of Human Life ? what its extreme limit ? should still be unsettled ; since for centuries there has existed concurrent testimony upon these points from writers of antiquity, both sacred and profane.

In the words of Jesus the Son of Sirach, ' The number of a man's days are at the most an hundred years,' [1] we have clear evidence as to what was been believed to be the limit of Human Life in his days ; and in the declaration of the Psalmist, ' The days of our age are three score years and ten, and though men be so strong that they come to fourscore years ; yet is their strength then but labour and sorrow, so soon passeth it away, and we are gone,' [2] we have unmistakable testimony as to what was then believed to be its average duration.

What was true as to the number of our days when those words were written the centuries which have since elapsed have not changed in the slightest degree ; and the most eminent physiologists of our own times are at

---

[1] Ecclesiasticus xviii. 9.   [2] Psalm xcv.

one with Ben Sira and the Psalmist in regarding three score years and ten as the average, and one hundred years as the extreme age of man.

And this latter is the view recognised by the Civil Law, clearly and distinctly laid down in these words, 'Vivere usque ad centum annos quilibet presumitur, nisi probatur mortuus.' ᴸ.' In the Civil Law,' says Taylor, to whose admirable work on the 'Law of Evidence'[2] I am indebted for this reference, 'the legal presumption of life ceases at the expiration of one hundred years from the date of the birth, and the same rule appears to have been adopted in Scotland, but in England no definite period has been conclusively fixed during which the presumption is allowed to prevail.'

That these propositions as to the average duration and extreme limits of human life are founded in truth, the researches of modern naturalists and physiologists abundantly testify.

What says Buffon? 'The man whose life is not cut short by accident or disease, reaches everywhere the age of ninety or one hundred years';[3] and he goes on to remark, which is very important—'If we consider that the European, the Negro, the Chinese, the American, the man highly civilised, the savage, the rich, the poor, the inhabitant of the city, and the dweller in the country, so different one from another in

---

[1] 'Corpus Juris Glossatum,' tome ii. p. 719, n. q.

[2] Vol. i. p. 189, ed. 1864.

[3] Buffon, vol. ii. p. 76. I quote from Flourens, 'De la Longévité Humaine,' 4th edition, Paris, 1860.

every other respect, agree in this one point, and have the same duration, the same interval of time to run through 'twixt the cradle and the grave, that the difference of race, of climate, of food, of comforts, makes no difference in the duration of life. . . it will at once be seen that the duration of life depends neither upon habits, nor customs, nor the quality of food, that nothing can change the fixed laws which regulate the number of our years.'

The fixed law which Buffon recognised was that in the animal economy, not in man only, the duration of life is regulated by the duration of growth. 'Man,' he says, 'grows in height until he is sixteen or eighteen, but his size is not fully developed until he is thirty. Dogs attain their full length during their first year, but it is only in the second they reach their full bulk. Man, which is fourteen years in growing, lives six or seven times that period, that is to say, till ninety or a hundred; while the horse, of which the growth is completed in four years, lives six or seven times that period, that is to say, from twenty-five to thirty years.'

Recognising the general accuracy of the principle laid down by Buffon, his distinguished countryman, M. Flourens, has sought to give greater precision to the law by settling the important question, what is the term or limit of growth. The great physiological pro-blem had been solved by Buffon's discovery that the duration of life depended on the duration of growth; all that remained was to ascertain of how many times the duration of growth the duration of life consisted.

The one thing which had escaped Buffon—namely, the
one certain sign of growth being completed, M. Flourens
claims to have found in the union of the bones with
their epiphyses.[1]   As long as the bones are not united
to their epiphyses the animal continues to grow ; but as
soon as such union takes place the animal ceases to
grow.   This union takes place in man at twenty, in the
camel at eight, in the horse at five, the ox at four, the
lion at four, and the dog at two ; and he then proceeds
to show how nearly accurate Buffon had been when he
said that every animal lives nearly six or seven times
the period of its growth—the truth being that it lives
about five.   Thus man being twenty years growing,
lives five times twenty—that is, one hundred years.

But I have yet one, and that a still higher authority
to produce.   Just as these sheets were put into the
hands of the printer, fortunately before they were composed, there appeared in ' Fraser's Magazine '[1] an article
' On Longevity,' from the pen of the most eminent
physiologist of the present day, Professor Owen, a paper
so important, instructive, exhaustive, and convincing,
that I entreat all my readers interested in centenarianism
to give it their most attentive consideration.   From the
essay, to which I shall have other occasions to refer,
and in which Professor Owen has done me the honour
to recognise my small services in the cause of scientific
truth, I extract the following passage, with which I
may fitly conclude my remarks on the average extreme
age to which man may attain.

[1] Flourens, ' Longévité, p. 85.
[2] ' Fraser's Magazine,' for February, 1872, pp. 218–233.

'The conclusion of Professor Flourens, that in the absence of all causes of disease, and under all conditions favourable to health and life, man might survive as long after the procreative period, ending—say at seventy in the male—as he had lived to acquire maturity and completion of ossification—say thirty years—are not unphysiological. Only under the circumstances under which the battle of life is fought, the possible term of one hundred years, inferred by Flourens as by Buffon, is the rare exception.'

After this, I need not occupy the time of my readers by adducing further proof that 'the number of a man's days are a hundred years at most.' But, as it is said, there is no rule without any exception, and certainly it is so in this case, though the exceptions are extremely rare, I feel bound, before passing away from this part of my subject, to notice two supposed exceptions to the law which fixes the limits of human life at one hundred years, which are very persistently brought forward with more or less show of reason.

The first of these is that the duration of human life was greater during the so-called 'good old times.' The second, that cases of centenarianism are found more frequently among the poor than among the rich, or, as Sir William Temple puts it, 'that health and long life are usually blessings of the poor, not of the rich, and the fruits of temperance rather than of luxury and excess.'[1]

It is true that if we could give credence to the in-

---

[1] Sir William Temple's Works, vol. iv. p. 339.

numerable stories of exceptional longevity which are recorded in Wanley's 'Wonders of the Little World,' and Easton's ' Human Longevity,' as having existed in this country, when ' Merry England,' was to a great extent covered with marshes and forests, our towns closely-packed and undrained, the hovels of the people ill-built and worse ventilated, their food coarse in quality and spare in quantity, and their clothes, with their accumu-lated dirt worn for years, it would be clear that an utter disregard for all sanitary laws would be the best security for the prolongation of human life. But we doubt whether even Dr. Wynter, who in his 'Essay on Longevity'[1] does not hesitate to question ' whether (in cases of longevity) the testimony of contemporaries is not of more value than a mere register of births and deaths,' would venture to supplement that paradox by the yet more startling one to which we have just directed attention. The fact is, those very parish registers, which Dr. Wynter holds so lightly, by furnishing, as they do when judiciously used, a very efficient means of correcting unfounded statements, have served materially to check the growth of such cases and diminish their number.

But I am enabled to produce some original evidence upon this subject, the value of which will be unhesi-tatingly admitted by all who share with me the ad-vantage of knowing Sir Thomas Duffus Hardy, and the authority with which he is entitled to speak on such a subject.

[1] 'Good Words,' July 1865, p. 493.

Knowing that he had investigated the question whether there was any, and, if so, what difference, in the average duration of life in this country in former times and the present day, I ventured to ask him what was the result of his inquiries, and in reply received from him the following letter :—

'Public Record Office, October 31, 1870.

'My dear Thoms,—Your letter of the 28th inst. has only just come to hand. It arrived here on Saturday after I had left the office. You are quite right in your remembrance of what I told you respecting my having some years ago looked into the subject of the longevity of men during the thirteenth, fourteenth, fifteenth, and sixteenth centuries, and I came to the conclusion that threescore years and ten was considered a great age, which few arrived at. I derived my data from the Inquisitions post-mortem, and the examination of witnesses in chancery suits, &c. The Inquisition gives the age of the heir, another the time of his death. We thus get at the age of the landed proprietor. The examination of witnesses in chancery and other suits gave the ages generally of country people. I don't say they are always to be relied upon. Another species of evidence is to be found in the Array Commissions, which call out all persons between the ages of 16 and 60, some 16 and 50. This shows that persons of the age of 60 were considered fit for fighting. If I can find my memoranda on the subject I will send them to you ; at any rate, you have my full authority to use my name to the effect that my belief, derived from our national records, is that persons, either in the higher or lower orders, seldom attained to the age of 80. I do not mean to say that they never attained that age, but I never met with an instance beyond that year that could be relied upon.   ·

'Ever sincerely yours,

'T. DUFFUS HARDY.

'Of course I mean during the period I have mentioned. Good drainage and other circumstances have, of course, tended to a greater longevity than formerly.'

'October 31. ·

'My dear Thoms,—In my letter just despatched I forgot to mention the case of Sir John Sully, who was examined as a witness when he was 105 years old, and Sir John Chydishe is said to have been examined when he was upwards of 100, but both cases are somewhat apocryphal, the latter especially.

'Very sincerely yours,

'T. DUFFUS HARDY.'

This seems so conclusively to negative the supposition that human life[1] in this country, centuries ago, was longer

---

[1] I am indebted to Dr. Sykes of Doncaster for calling my attention to the following passage on the length of man's days, written by one of the most popular English divines of the fourteenth century, Richard Rolle de Hampole, who, in his 'Pricke of Conscience,' bk. i., ll. 728-63 (I quote from Dr. Morris's admirable edition) writes as follows :—

> 'In þe first begynnyng of þe kind of man
> Neghen hundreth wynter man lyfed þan,
> As clerkes in bukes bers witnes ;
> Bot sythen becom mans lyf les

than it is now, that I pass to the consideration of the second supposed exception to the general law, namely, that the duration of human life is greater among the poor than among the rich.

One of the most satisfactory explanations of the popular error, for such it unquestionably is, that Longevity is most frequent among those most exposed to privations and hardships, is that so quaintly described by Fuller in his ' Holy War' (chap. xix. of Supplement) :

And swa wald God at it suld be ;
For whi he sayd þus til Noe ;

Non permanebit spiritus meus in homine in eternum, quia caro est; erunt dies illius centum viginti annorum.

" My gast," he says, " sal noght ay dwelle
In man, for he is flesshe and felle ;
Hys days sal be for to life here
An hundreth and twentie yhere."
Bot swa grete elde may nane now bere,
For sythen mans lyfe bycom shortere.
For-whi þe complection of ilk man
Was sythen febler þan it was þan ;
Now is it alther-feblest to se,
þarfer mans life shert byhovest be ;
For ay þe langer þat man may lyfe
þe mare his lyfe sal hym now griefe,
And þe les him sal thynk his lyf swete,
Als in a psalme says the prophete :

Si autem in potentatibus octogynta anni, et amplius eorum labor et dolor.

" If in myghtfulnes four score yhere falle,
Mare es bair swynk and sorow with-alle,"
For seldom a man þat has þat held
Hele has, and himself may weld ;
But now fallen yhit shorter mans dayes,
Als Job, þe haly man, þus says :

Nunc paucitas dierum meorum finietur brevi.

" Now," he says, " my fou days sere
Sal enden with a short tyme here." '

'Armies both of Europe and Asia (chiefly the latter)
are reported far greater than truth.   Even as many old
men used *to set the clock of their age too fast* when once
past seventy, and, growing ten years in a twelvemonth,
are presently fourscore ; yea, within a year or two after,
climb up to a hundred.'

This tendency to ' set the clock of their age too fast '
is common to old people of all classes alike ; but in the
higher ranks it is at once corrected by family evidence,
the records of the Heralds' College, and similar sources
of information, but which leave the self-delusions of the
village Hampdens and mute inglorious Miltons un-
checked and uncorrected.   That in spite of poverty, toil,
hard fare and exposure, more in number (*not* in proportion) of the humbler classes become Centenarians
cannot be doubted, but the reason is a very obvious
one, namely, that whereas a very limited per-centage of
people ever attain to or exceed the age of a hundred,
the poor being to the rich as millions to tens of
thousands, Centenarianism in the humbler classes pre-
ponderates in the same rate over Centenarianism in high
places.

How rare this latter is, few who have not examined
the question can imagine.   ' It may, I believe,' said Sir
George Lewis,[1] ' be stated as a fact that, limiting ourselves to the time since the Christian era, no person of
royal or noble rank mentioned in history, whose birth
was recorded at the time of its occurrence, reached the

. [1] 'Notes and Queries,' (April 12, 1862, 3rd S. i. 81.) See also his letter
to Mr. Twisleton, under date May 10, 'Correspondence,' p. 416.

age of a hundred years. I am not aware that the modern peerage and baronetage books contain any such case resting upon authentic evidence.'

But the reader may argue that this only serves to prove that the duration in this apparently favoured class is below the average ; but that is not so. In an elaborate paper 'On the Rate of Mortality prevailing amongst the Families of the Peerage during the Nineteenth century,'[1] by Mr. Arthur Hutcheson Bailey, and Mr. Archibald Day, the well-known actuaries, these gentlemen distinctly show 'that the average mean duration of life among the families of the peerage is throughout *materially greater* than with the general population.' And yet in the face of this startling fact, I believe no authenticated case of Centenarianism has ever occurred in any noble family.

Since I commenced the investigations which have resulted in the present work, two such instances have been brought before me. The first was that of the old Marquis of Winchester (*Salix non Quercus*), said to have died at the advanced age of 106 ; the second that of Lady Mary Bouldby, who, as I was confidently assured, had also reached the same extraordinary age.

Of the former it is said in Wanley's 'Wonders of the Little World,'[2] on the authority of Baker's Chronicle, ' He lived in all an hundred and six years and three-quarters and odd days, during the reigns of nine kings and queens of England.' But the more careful and

---

[1] 'Journal of the Statistical Society of London,' March 1863, pp. 49–71.
[2] Vol. i. p. 91, ed. 1806.

trustworthy Camden tells us in his 'Annals' under the year 1572. 'This year a peaceable death took away William Powlett, Lord High Treasurer of England, Earl of Wiltshire, and Lord St. John of Basing; a man that had passed through very great honours. He died in the *ninety-seventh year* of his age, after he had seen' one hundred and three persons that were descended from him.'

Lady Mary Bouldby's supposed extreme age was unsupported by any evidence, and her 106 years, like those of the Marquis of Winchester, proved upon investigation to be greatly exaggerated. She was the second daughter of George, third Earl of Cardigan, who succeeded his grandfather, July 16, 1703, and died July 5, 1732, leaving four sons and two daughters.[1] Lady Mary, the younger of these, was first married to Richard Powys of Hindesham in Sussex, by whom she had two daughters, and secondly, on June 2, 1754, to Thomas Bouldby of Durham, and died on February 21, 1813, as the prosaic 'Annual Register' tells us 'aged 97.'

But if like Sir George Lewis I have failed in finding any well-authenticated case of Centenarianism in the peerage, the baronetage has proved slightly more productive. Catherine, the third daughter, and one of the twelve children of Sir John Eden, Bart., of Windleston, was born on February 10, 1771, and baptised on the following day in the church of St. Andrew, Auckland. In 1803 she married Mr. Robert Eden Duncombe Shafto, of Whitworth Park, Durham, and died on March 19,

---

[1] Collins's 'Peerage' (by Brydges), iii. 497.

1872, having more than completed her hundred and first year. This case admits of no doubt, the lady having in 1790, when nineteen years of age been selected as one of the Government nominees in the Tontine of that year, as will be seen by the following interesting communication on the subject, with which I have been favoured by Sir Alexander Spearman :—

'The Spring, Hanwell.

'My dear Sir,—I send you a mem. relating to Mrs. Shafto, and another Tontine case, which it may be satisfactory. to you to see.

'Yours very truly,

'A. SPEARMAN.'

'Thursday, April 18, 1872.'

'National Debt Office.

'Mrs. Shafto, daughter of Sir John Eden, Bart., of St. Andrew, Auckland, Durham, appears by the records of the Tontine, 1789, to have been nineteen years of age in October 1790, when the nominees were nominated. Neither she nor her family had any interest in the Tontine. She was one of the nominees selected by the Lords of the Treasury to keep up the full number of lives.

'The lives selected by the Treasury were the children of peers, baronets, lords of manor, &c., being of that class in which the parties selected could easily be kept in view in future years. Mrs. Shafto's life and identity have been proved, first to the Exchequer, and since 1832 to the Commissioners continuously up to the present year, when she died. There is certainly no

reason to doubt that she died in the hundred and second
year of her life. The nature of the evidence of age and
the mode of selection of these nominees will be seen by
reference to Mr. Tomlinson's report on Mortality of An-
nuitants, printed as a House of Commons paper (585,
September, 1860). Nothing is known at this office with
respect to the views said.to have been entertained by
Sir Cornewall Lewis on the subject of Longevity. He
had no special information given him on the point from
here. The experience of this office shows certainly one
other life that lasted over 102 years, viz., David Rennie
of Dundee, farmer, who died on March 2, 1857, having
been born February 28, 1755. The evidence in this
case was perfect.'

The importance of this statement as to the experience
of the National Debt Office with respect to the question
of Human Longevity, made as it is on the high authority
of Sir Alexander Spearman, cannot be over estimated ;
more especially when it is remembered that we have in
it the result of an experience, not on lives taken at
random, but on a series of lives selected in the belief
that they will prove to be of long continuance.

The experience of the National Debt Office is, as
might be expected, confirmed by that of the numerous
Assurance Offices.

As long since as January 31, 1857, there appeared in
the 'Athenæum' an article on Longevity by Dr.
Webster, an earnest inquirer after truth, but disposed to
believe in the more frequent occurrence of Centenarianism than I believe to be justified by experience. In this

paper there occurs the following passage, in reply to a challenge which had been thrown out to him in reference to a case under discussion, to produce a fact equally marvellous from the records of Life Assurance :—

'Respecting Life Assurance, the question is, however, more easily answered, as shown from the following reports from twelve of the largest and longest-established offices in London, who kindly supplied every requisite information in regard to the deaths of parties insured, or very extreme ages still existing. At the Amicable, the most aged on whom a policy had been paid died at 97 ; the Pelican, 97 ; Royal Exchange, 96 ; Equitable, 95 ; Albion, 95 ; Rock, 94 ; Imperial, 94 ; Union, 94 ; Atlas, 92 ; Law, 92 ; Sun, 92 ; and London, 90. Besides the above facts, it may be remarked that, at several Companies designated, various persons whose lives are insured and still live, have attained equally advanced ages, although none have yet become Centenarians as far as I could ascertain.'

Dr. Webster proceeds to argue against the value of this evidence on several grounds—one being that the majority of Centenarians are found among the poorer classes, who do not insure their lives, an assumption for which I do not believe there is the slightest foundation. Another objection which he takes is that 'the adoption of life insurance upon an extensive scale is only of modern date'; but, as was well remarked by Mr. Dilke in reply, 'The Amicable Life Insurance Office was established in the reign of Queen Anne . . . that

taking Dr. Webster's assumption that assurances on
lives are mostly entered into at forty years of age, the
first insurers must have been born as early as 1670, so
that Dr. Webster's inquiries establish the fact that from
1670 to 1857, "*no solitary instance* has occurred of a
person who had assured his life attaining a greater age
than 97."'

In the fifteen years which have since elapsed (1857 to
1871 inclusive), the inference which Dr. Webster was
inclined to draw that some of the 'parties of equally
advanced age,' still living in 1857, would become Centenarians, has not been justified. For what is the
evidence as to Centenarianism furnished by Assurance
Offices up to the present time. This is shown in the
accompanying letter from my friend Mr. Bailey, of the
London Assurance Corporation, to whom, from my
desire that statements of such importance should rest
upon higher authority than I can lay claim to in such
matters, I applied for information :—

'*The London Assurance Corporation*, 7 *Royal Exchange,*
London, E.C. Life Department.

'April 9, 1872.

'Dear Sir,—I have made several inquiries at your
request, in order to ascertain with accuracy the oldest
ages that have been attained by persons whose lives
have been assured. The result is that in the entire
experience of the Life Assurance Companies of this
country there has been but one case of a Centenarian,
that of Mr. Luning, the particulars of which have been
published, and are well known to you. I have met with

one death in the hundredth, and three in the ninety-ninth, year of age. Life Assurance Societies have existed in this country since the year 1706, and as for their own purposes the mortality experience of the principal offices has at different times been ascertained by elaborate investigations, I think that this indirect evidence on the subject of Centenarianism is of some value.

'I am, dear Sir, yours very truly,

'K. H. BAILEY.'

'To W. J. Thoms, Esq., 40 St. George's Square, S.W.'

If so many thousand selected lives of well-to-do people, whose correct ages can be clearly established, give such a very small per-centage of Centenarians, it surely justifies us in regarding the belief that poverty is favourable to Longevity as a popular error, and receiving with caution the numerous cases of Centenarianism brought forward without a particle of evidence in support of them, and of which, more abundant even than Byron heroes,—

'Every day and week sends forth a new one.'

CHAPTER III.

BEFORE proceeding to consider the nature of the evidence on which cases of abnormal Longevity can be satisfactorily established, it is necessary, strange as it may seem, to show by whom such evidence should be produced. For the general practice is to assert that Old So-and-so is of some exceptional age, and to call upon those who doubt it to disprove the statement.

I may therefore be pardoned if I again insist upon this one point—too often overlooked—that it is the duty of those who bring forward instances of alleged Centenarianism to accompany them with the evidence necessary to establish their truth, and *not* to call upon those who doubt such unsupported statements to refute them.

Common sense and the rule of the Civil Law: 'Ei incumbit Probatio qui dicit non qui negat,'[1] alike call for this; and not only for this, but in proportion as the Centenarian is stated to have exceeded the normal life of man, that the proof of it should be clear, distinct, and beyond dispute, or as Coke puts it, 'Proofs ought to be evident, to wit, clear, and easily understood.'[2]

---

[1] 'Starkie on Evidence,' i. 418, who also quotes from Justinian, 'Probandi necessitas incumbit illi qui agit.'—*Inst.* lib. 2, tit. 20.

[2] 'Probationes debent esse evidentes, scilicet, perspicuæ et faciles intelligi.'—*Co. Lit.* 283.

Upon asking one of the most distinguished lawyers of the present day, who, to use Lord Campbell's pet phrase, has 'himself filled the marble chair,' if he could kindly refer me to any rule of law upon this point, or to any case in which this rule had been laid down, he said, No, it was so obvious that no such dictum or ruling existed ; that he himself had always laid down that the amount of evidence necessary to be produced in a case depended entirely upon the antecedent circumstances. When those circumstances are probable and consistent with ordinary experience, a very small amount of evidence will suffice to establish them. But if they are exceptional and improbable, just in such proportion must the evidence be clear, distinct, and irrefragable.

Having as I trust proved by whom the evidence should be produced and its nature, I will make a few remarks on the chief species of evidence usually brought forward in these cases.

These are five in number : 1. Baptismal certificates ; 2. Tombstone inscriptions ; 3. The number of the Centenarian's descendants ; 4. The recollections of the Centenarian ; and, 5. The evidence of old people still living, who knew the Centenarian as 'very old' when they themselves were quite young.

Of these various species of evidence there is none so universally considered to be beyond dispute as a certificate of baptism. It is therefore clearly desirable to call attention to the caution with which such evidence ought to be received.

At first sight, it would seem that nothing could be

more direct and satisfactory than an official certificate
showing when and where the alleged Centenarian had
been baptized; the registration being contemporary and
the date precise. But when a doubt is suggested as to
what proof there is that this is a certificate of the baptism of the individual into whose age we are examining,
it is at once seen how unsatisfactory is the evidence
afforded by such certificate, unless supported by corroborative facts. All who were present at the baptism—
the sponsors who held the child at the font, the priest
who administered the holy rite—every individual who
could have borne testimony to the identity of the supposed Centenarian with the individual named in the
register of baptisms, have long since passed away, and
nothing is left but to trust to secondary and circumstantial evidence.

Fortunately, in a large proportion of cases, little
difficulty will be found in obtaining such corroborative
evidence from facts connected with the relations which
the Centenarian bore to the members of his family, his
occupation, mode of living, &c. The dates of birth, surname and Christian name of his father and mother, the
place and date of their marriage, the birth dates of any
brothers or sisters he may have had, the date of his own
marriage and of the births of his children, of his admission into the school at which he was educated, his
entrance into the army, navy, or any other public employment, his apprenticeship, all furnish points for
inquiry tending to elicit information as to his identity
and real age.

For obvious reasons, as much information as possible upon these points should be obtained before proceeding to search for the register of baptism, so that the inquirer may not be misled into taking an entry of the baptism of a child of the same name as the Centenarian, but possibly the issue of different parents, for the evidence of which he is in search.

The necessity for such precaution will be seen from the following very interesting case.

On December 20, 1863, a maiden lady died at Liverpool of such exceptionally great age that her medical attendant felt justified in calling special attention to her case in 'The Times,' where her death was announced in the following terms :—

'December 20th, 1863, at her residence, Edge Lane, Liverpool, aged 112 years and 6 months, Miss Mary Billinge. She was born at Eccleston near Prescot on the 24th May, 1751. She retained her faculties in a very remarkable degree to the last, and was never known to have been confined to her bed for a single day until the week preceding her decease.'

In January 1865, at which time the question of Longevity was under discussion in 'The Times,' Mr. Newton, the professional gentleman alluded to, again brought forward in that journal the case of Miss Billinge ; when, both in 'The Times' and in 'Notes and Queries' (3rd Ser. vii. 154), he was pressed to give satisfactory proof of the identity of the deceased lady with the child baptized in 1751.

In reply to this challenge, Mr. Newton stated that

great difficulty had been experienced even in ascertaining that Miss Billinge had been born at Eccleston near
Prescot, that the matter had been investigated by the
authority of the Health Committee of Liverpool, that
the officer authorised to make inquiries had, after some
research, 'rested quite satisfied with the truth of the
certificate.' Mr. Newton mentioned, in addition, two
little facts which eventually proved to be very important, namely, that Miss Billinge had a brother and
sister, and was the senior of both, and that the brother
died in 1817, aged forty-seven years.

This latter fact showing that her brother was nineteen years younger than herself did not tend to increase my belief that Miss Billinge was upwards of
112 at the time of her death; but as I was unable to
go to Liverpool to investigate the case I was content
to wait till I could interest some friend there in the
inquiry.

This I was eventually enabled to accomplish. I
cannot do better than give the result in the words of
the kind and intelligent friend who came to my assistance, and reported the case in the columns of 'Notes
and Queries' (3rd Ser. vii. p. 503).

'I am now in a condition to furnish satisfactory
information on the subject of the age of the supposed
Centenarian, Miss Billinge;[1] and I will in a few words
describe the process by which I have arrived at it.

[1] The supposed 112 years of Mary Billinge are destined, I fear, to exercise for a long time an unfounded influence on the question of the duration
of human life. They are duly recorded in 'the 26th Report of the Registrar-
General of Births, Deaths, and Marriages;' while in a thoughtful and well

'On application to Mr. Newton, surgeon, I was furnished with a copy of the certificate of baptism of "Mary, daughter of William Billinge, farmer, and Lidia his wife ; born 24th May, 1751, and christened the 5th of June." This was assumed to be the Mary Billinge recently deceased. The question thus became one of identity. After some inquiry, I found Miss Billinge had a brother and sister buried in Everton churchyard. I have extracted the inscriptions on their tombstones as follows :—

"William Billinge, obt. 7th May, 1817, aged 46.
Anne Billinge, died 9th Feby., 1832, aged 59."

'I have also seen a mourning ring which belonged to the late Miss Billinge, in memory of her brother, which confirms the above date of his death. It is clear, therefore, that William and Anne were the brother and sister of the late Mary Billinge.

'The next point was to ascertain the parentage of William and Anne. I went over to Prescot church, and found the parish clerk—himself a relic of antiquity, ninety years of age, and still doing duty. He made a search for me, and found the registers of both :—

'William in 1771, son of Charles and Margaret Billinge.

'Anne in 1773, daughter of the same.

written book, published in 1865, entitled 'Man's Age in the World,' by an Essex Rector, we read : 'The utmost modern powers of man *authenticated*, may be placed thus—

Thomas Parr .	.	.	.	.	A.D. 1635,	aged	152
Henry Jenkins	.	.	.	.	,, 1670	,,	169
Mary Billinge .	.	.	.	.	,, 1863	,,	112
Sarah Lee	.	.	.	.	,, 1864	,,	105.'

'It was clear then that William and Anne, children of Charles and Margaret Billinge, could not be brother and sister of Mary, the daughter of William and Lidia Billinge.

'To put the matter beyond a doubt, I persevered in the search, and found :

" Mary, daughter of Charles and Margaret Billinge, born 6th November, 1772, christened 23rd December."

'The identity is here complete. The old lady was, therefore, in her ninety-first year,[1] not in her 112th when she died. I suspect that most of the supposed instances of Centenarianism will turn out to be cases of mistaken identity.

'J. A. P.'

After such a striking proof of the manner in which intelligent inquirers, animated by a sincere desire to ascertain the truth, may be misled in investigations of this nature, it is surely no unreasonable law to lay down that certificates of baptism, unsupported by corroborative testimony, cannot be received as evidence of Longevity.

But there is yet another caution to be observed in the use of these registers. Even when the names of the parents have been ascertained, and a child of such parents and such name is found on the register, it does not necessarily follow that such entry refers to the supposed Centenarian. Parents, anxious to perpetuate a particular family Christian name, will frequently, on the

---

[1] Or rather, as Mr. Newton afterwards pointed out, she was really 91 years, 1 month and 14 days.

death of a young child, to whom it has been given,
bestow it on a second child, and in the event of the
death of that child, christen a third child by the same
name. How frequently this happens the following
incident will show. At a small party of not more
than twelve or fourteen, at which I was present, I
was the subject of some good-natured quizzing for
my scepticism with regard to a case of Centenarianism ; the only evidence, by which it was at-
tempted to support it being a baptismal certificate.
Having justified my doubt, on the ground that the
child referred to might have died, and that the Centenarian would probably be found to be a younger
brother, the reasonableness of my doubt was at once
maintained by a distinguished Royal Academician who
was present, and who stated that he was the *third son*
of his parents who had received the same Christian
name. Two infant brothers who had borne his Christian
name had predeceased his birth, and in the parental
anxiety to perpetuate the name, it had been bestowed
upon him, their third boy. At this same party, a
medical friend mentioned, in confirmation of this practice, that his wife was the third daughter of her father
and mother, who had borne the same Christian name,
and I may add that the friend who cleared up for
me the mystery of Mary Billinge bears the same two
Christian names as an elder brother who had died an
infant.

As, with few exceptions, reputed Centenarians are
found among the poorer class—a class which from not

possessing any papers, frequently not even a family Bible, renders the thorough identification of an individual belonging to it a matter of great difficulty ; and as the only evidence produced in these cases, is, as a rule, the register of baptism, a document of the highest value when properly established, but otherwise one of the most fallible of tests,[1]—I am desirous of pointing out some of the precautions to be observed in searching for and accepting such documents in the examination of cases of supposed exceptional Longevity.

I am the more anxious to insist upon this, since in the course of an extensive correspondence, which I have lately carried on in investigating a number of cases of supposed Centenarians, I find that a very erroneous impression exists as to the value of this evidence, *per se*, and as to the necessity of having it corroborated by other proof.

In the investigation of all cases of abnormal longevity, I would urge the inquirer, before proceeding to search a register for a certificate of the baptism of the supposed Centenarian, to ascertain from him some of those facts which serve to point out which, among many entries possibly relating to the same family, at all events to the same name, is the entry required. For this purpose it is expedient, without wounding the susceptibilities of the old person, naturally sensitive on the subject of his

---

[1] I have reason to believe that the Civil Service Commissioners have in more than one instance had occasion to call for proofs of the identity of a candidate who had presented himself for examination, with the baptismal certificate, produced as evidence of his age ; and that the result has ustified the precaution.

age, so to lead the conversation as to learn what was the Christian as well as the surname, both of his father and mother, more particularly the latter ; and, if possible, where and when they were married ; whether he had any brothers or sisters, and, if so, their names, dates of birth, &c.; whether he was ever in the army, navy, or other public employment, and, if so, at what dates ; if married, the date and place of his marriage, and the names and dates of birth of any children born of such marriage.

Thus furnished with the means of identifying the person the date of whose baptism is the object of his search, the inquirer is not likely to fall into the common error of supposing the first baptismal certificate he meets with of a John Nokes to be necessarily that of the John Nokes who is the subject of investigation, but will pursue his researches until at length he discovers the baptism of the true John Nokes, who generally proves to be some ten or fifteen years younger than the world had been led to believe.

And here let me again warn inquirers about a source of error which, as I have shown, is more frequent than is generally supposed. Searching, let me say, for the certificate of John Nokes, who died in 1870 at the reputed age of 105, the inquirer, having procured all the preliminary information I have indicated, turns to the register for 1765, and there, sure enough, he finds the baptism of John Nokes, the son of John and Mary, and at first believes that he has proved that old John Nokes was right, after all, and had really attained the ripe age of 105. Still, to be quite safe, the inquirer turns over a few more leaves of the register, and lo ! in 1770 he finds the entry repeated—another John Nokes, the son of John and Mary. He is puzzled at this, but the burial register clears up the mystery. John Nokes *primus* died young, and his parents christened another son after his father, John. So old John Nokes after all was only 100. But the inquirer, taught caution by experience, turns once more to the burial register, and there finds that Nokes *secundus* had been laid by the side of Nokes *primus*; and a few years later he finds the baptism of the third John Nokes, the old man who died in 1870, not five years over the hundred, but a few years under.

The practice of giving the same Christian name to successive children, by way of perpetuating it is very frequent, and I some time since was furnished by another correspondent from Guernsey, with a curious illustration of it. 'A friend,' says my correspondent, 'with whom I was this morning talking the matter over, named a case in this island in which a gentleman had three sons, named John Peter So and So. My friend was himself misled as to the age of the John Peter with whom he had business, believing him, on the authority of the parochial register, to be much older than he really was. Eventually it was explained that he had in his researches taken John Peter the first for John Peter the third.'[1]

---

[1] The same writer called my attention to the case of Gibbon, the historian but in his case the motive for repeating the Christian name was

Another friend has sent me an extract from the pedigree of Henry Hibbert of Preston, East Yorkshire, which shows that by his marriage with Anne Burrell, he had—1. a son HENRY, born July 2, and buried July 10, 1660; 2. a son HENRY, born October 14, 1661, buried August 18, 1665; and 3. a son HENRY, born January 20, 1672–3, and buried March 16, 1679–80.

Professor Owen, in the valuable paper to which I have already referred, points out another source of error by which the most careful inquirer is liable to be misled. 'The system of registration of births,' says Professor Owen, 'now affords the competent searcher after truth the needful date. Parish registers cover a greater period of time. But both have their sources of fallacy, needing caution. In the case, let us say of Richard Roe, reported now living at St. Hilda's, North-shire, at the extraordinary age of 120: one writes to the incumbent, respectfully requesting that a search may be made in the vestry for any evidence of said Richard's birth. An answer is duly received that Richard Roe was baptized in November 1751. This seems straightforward and satisfactory. But the worthy incumbent is again troubled with the request that the parish register may be further searched for the birth or baptism of any other Richard Roe at a later period;

anticipatory of his possible death. 'So feeble was my constitution,' he says in his Autobiography, 'so precarious was my life, that in the baptism of my brothers, my father's prudence successively repeated my Christian name of Edward, that in case of the departure of the eldest son, the patronymic appellation might be still perpetuated in the family: "Uno avulso non deficit alter."'

and for the entry of the marriage, if perchance such
may have occurred in the place of the birth of the
Richard Roe of 1751, and of that of any subsequent
Richard Roe.

'It turns out that the Richard Roe of 1751 married
early, viz. in 1769, one Margaret or Margery Doe, of
the same parish, and that their eldest born was christened Richard Roe in 1771. Now this Richard Roe, it
further appears, wedded at St. Hilda's Elizabeth Bunch,
of the same parish; and surviving Bunches know well
that such was the name, and not Doe, of the first wife
of the wonderful old man. Whereupon it appears that
the Richard Roe in question has reached his 100th, not
his 120th, year.'

Before quitting the subject of parish registers let
me again urge upon intending inquirers how desirable it is to ascertain, before proceeding to search
these documents, the particulars from which the corroborative proof of identity may hereafter be obtained;
and illustrate by a few examples how satisfactory the
plan of eliciting preliminary information has already
proved.

I have instanced the case of Mary Billinge, whose
identity was established by ascertaining the parentage
of her brother and sister. An old fellow, William
Webb, who died recently, long supposed to be considerably over a hundred, on the strength of a baptismal
certificate, was eventually ascertained to be considerably
younger when the Christian name of his mother was
ascertained. Webb's real age, some ten years less than

was claimed for him, was confirmed by the records at the Admiralty, which records, in like manner, reduced the age of Edward Couch supposed to be 110, to the more ordinary age of 95.

In the same way the books of the Admiralty, the War Office, and Chelsea Hospital, have corrected the ages of several claimants to Centenarian honours, such as Thomas Geeran, the Brighton Centenarian, the so-called 'Captain' Lahrbusch, Joseph Miller, and many others whose cases will be detailed hereafter, and will I think justify the pertinacity with which I urge these suggestions, in the hope and belief that they will prove of use to those who, taking an interest in the great question of the Duration of Human Life, hold with me that on this particular point truth, if not stranger than fiction, is, at all events, a great deal better.

CHAPTER IV.

NEXT to baptismal certificates, tombstones are by many regarded as evidence not to be resisted in cases of supposed abnormal Longevity. But the experience and the records of our courts of law alike serve to show that not only are epitaphs and monumental inscriptions to be distrusted for the fulsome and ûnmerited eulogies which they too often recite ; but also that the very facts and dates which they profess to record are often not to be depended upon.

These errors of date are sometimes the result of mere carelessness or ignorance ; sometimes they have been made deliberately for the gratification of personal vanity ; sometimes for the baser purpose of falsifying pedigrees or bolstering up fraudulent claims to titles and estates.[1]

From this it will be seen that an inscription on a tombstone, unless supported by corroborative evidence,

---

[1] In the celebrated claim to the Tracy Peerage, which was heard before the Committee for Privileges in the House of Lords, in the year 1848, it was sought to establish one important link in the chain of evidence by the inscription on an old tombstone which had been found in the churchyard of Castlebrock. The evidence was, however, not received for the very sufficient reason that the Crown called as a witness—the man who had cut the inscription. Lord Brougham put the credibility of the witness to the test by making him then and there cut some further words on the same stone.

cannot be received as proof of the fact stated, whether it be the age of a reputed Centenarian, or any other particulars of the party commemorated by it.

Thus, it has been shown by Colonel Chester in a very interesting paper read before the Historical Society, and entitled 'An Official Inaccuracy respecting the Death and Burial of the Princess Mary, Daughter of King James I.'—that while the Princess really died on September 16, 1607, both the official register of Westminster Abbey, and her monument in Henry VII.'s Chapel, state that her death took place three months later, viz. December 16, 1607. Colonel Chester's explanation of the source of this error is very ingenious; but with that I must not trouble my readers, as my only object is accomplished by showing that even the facts and dates stated on a royal monument cannot always be relied upon.

Again, many of the extraordinary ages recorded on tombstones are supposed to be the result of ignorance, as in the case at Chave Priory, in Worcestershire, 'where there is one which,' says the 'Quarterly Review' (vol. cxxiv. p. 181), 'ascribes to an old forefather of the hamlet the goodly length of 309 years. But the record meant nothing patriarchal. The village chiseller, hazy about numeration, wished to score 39, and engraved 30 first and 9 afterwards.' In the churchyard of Bickenhill, Warwickshire, is a tombstone to the memory of a Mrs. Ann Smith, who died in 1701. It states that she 'died a maid, and deceased, aged 708!' Whether that is the notation of some Warwickshire stone-cutter for

78, or the 8 has been added wantonly, does not appear. At Stratford-on-Avon, some workmen, engaged in the restoration of the church about the year 1839, having found a gravestone in which there happened to be a space before the age 72, for the honour of the place, and it is suspected with the consent of the sexton, inserted the figure 1 in the space, and so changed the 72 into 172.[1]

The following amusing instance of the credulity of parish authorities, and of their anxiety to perpetuate the memory of a supposed remarkable inhabitant of the village (for which I am indebted to Mr. J. O. Phillipps), is recorded by Warner, in his 'Tour through the Northern Counties of England' (1802) vol. i. p. 11:

'On passing through Brislington, two miles from Bristol, we could not help smiling at an instance of modern credulity which an inscription on an ancient stone in the churchyard hands down to posterity. About thirty years ago, the active churchwardens of Brislington, in clearing the churchyard and its accompaniments, discovered on an old tomb the notification of a remarkable instance of Longevity: " 1542.   Thomas Newman, aged 153." With due regard to the preservation of so curious a fact, they had the tomb repaired and brushed up, and the following inscription added to the original one: " This stone was new faced in the year 1771, to perpetuate the great age of the deceased." It was not till their official authority to repair and beautify, pull down and remove, had ceased, that they understood the figure 1 had been prefixed by a wicked

---

[1] 'Notes and Queries,' 1st series, viii. 124.

wit ; and themselves duped by this false addition, which
gave an antediluvian age to an honest man who died
before he had reached his grand climacteric !'

A correspondent in 'Notes and Queries' (4ᵗʰ S. iii.
593), lately called attention to the following inscription,
which occurs on a tombstone in Fulham Churchyard :—

'Under this Stone
are deposited the Remains of
NATHANIEL REUCH,
late of this Parish, Gardener,
who departed this Transitory Life
January 18th, 1783,
Aged 101 Years,' &c.

But the intelligent writer wisely adds :—' On turning to
Lysons' "Environs" (edit. 1795, ii. 375), I find that
when the matter came to be inquired into, the said
Nathaniel Reuch was proved to be born in the month of
August, 1701, and was thus 82, and not 101 years of
age at his death.'

This was doubtless an error, occasioned by receiving
without question the statement of the supposed Centenarian ; but it is not very easy to •account for the
following discrepancy :

Charles Macklin, the celebrated actor, is supposed, on
the authority of the monumental tablet erected to his
memory in the Church of St. Paul's, Covent Garden, to
have reached the exceptional age of 107. This does
not tally with the account given of him in the ' European Magazine,' vol. xxxii. p. 317, where it is stated
that ' his death happened on July 11, 1797, at the great
age, it is supposed, of 96 years.' But in spite of this,

thanks no doubt to the inscribed tablet, Macklin figures as a Centenarian, not only in all books relating to Longevity, such as the works of Easton and Bailey, but in Cates's excellent 'Dictionary of General Biography,' where we read 'he died in 1797 at the patriarchal age of 107.'

But he was not 107. About the year 1859 the vestry of St. Paul's, Covent Garden, came to a resolution, consequent upon the closing of the graveyard, to cover up the coffins laid in the vaults. In carrying out this resolution, the churchwardens came upon the coffins of Macklin and his wife—who, be it remarked, outlived him ten years, being only 64 when Macklin died. She furnished most probably the particulars inscribed upon his coffin-plate. They are as follows :—

> MR. CHARLES MACKLIN,
> Comedian,
> Died 11 July,
> 1797,
> Aged 97 Years.

So the name of Charles Macklin must be erased from the roll of Centenarians.

After the instances I have thus given, and they might easily be multiplied,[1] my readers will I think be disposed to agree with me, that in cases of supposed abnormal Longevity, the evidence of age furnished by a monumental inscription, can only be admitted when supported by corroborative facts.

Before quitting this part of my subject, I am bound

---

[1] See the latter part of this volume for particulars of the monuments of Henry Jenkins and Betty Evans.

E

to explain that the errors on monuments are not all of the same character. For instance, in 'The Times' of the 7th January, 1865, appears a letter from 'A Man of Kent,' in which the writer says : 'The inscriptions on some mural slabs to the memory of several of my ancestors in a parish church in this county have lately been restored; and on a recent visit to inspect the restoration imagine my feelings on finding it recorded as a fact, that a man, whose fourth wife survived him, departed this life in the 11th year of his age ; 61 had been converted into 11 !'

Another species of evidence is often brought forward in support of cases of alleged Centenarianism, in a manner which shows that those by whom it is produced look upon it as irresistible. A little consideration, however, will I think show that it is of comparatively small weight.

I allude to the manner in which it is sought to establish the fact that old Mrs. Smith must be more than 100 years old, because she has so many children, grandchildren, and great-grand-children still living.

The number of living *descendants* proves nothing ; the number of *existing generations* is another thing, and would be far better evidence, for there may be many descendants in few generations, and though the number of generations may be remarkable, the number of descendants belonging to each generation may be limited ; but even the number of generations does not necessarily establish what it is brought forward to prove.

Let me illustrate this by an imaginary case. It is

exceptional I admit; but not a whit more than the
101, 2, or 3, claimed for old Mrs. Smith :—

Mary Jones marries John Smith when she is
quite young; and (the case is by no means un‑
common), Mrs. Smith becomes a MOTHER at
eighteen   18

Her eldest child, a daughter, proves to be equally
precocious, and gives birth to her first child when
she too is only 18 ; Mrs. Smith becomes a GRAND-
MOTHER when she is *thirty-six*   36

This grandchild, also a female, marries young,
and becomes a mother at 18, and so old Mrs. Smith
becomes a GREAT GRANDMOTHER at *fifty-four*   . 54

The example is followed in another generation,
and the result is that the great grandmother of 54
becomes a GREAT GREAT GRANDMOTHER at
seventy-two   72

The child then born being as supposed, in all
preceding cases, a daughter, and exhibiting the
family precocity, marries early, becomes a mother
at 18, and old Mrs. Smith, wanting ten years of
the century, becomes, what I believe is very rarely
seen, a GREAT, GREAT GREAT, GRANDMOTHER
at *ninety*   90

That is to say, ten years before old Mrs. Smith arrives
at the distinction of Centenarianism.

Yet though great grandchildren, who are the third
generation, are frequently mentioned, and great great
grandchildren, who are the fourth, are occasionally mentioned, I do not call to mind a single instance in which

anomalous Longevity has been sought to be made credible by the averment of the existence of the descendants of the fifth generation, that is to say, of great great great grandchildren.

In the very remarkable and clearly established case of Mrs. Williams of Bridehead, County Dorset, who having been born on November 13, 1739, was at the time of her death, October 8, 1841, within a month of completing her 102nd year, there is no mention made of any descendants beyond grandchildren and great grandchildren.

But it may be urged that I am drawing extreme inferences from very exceptional premises. But are these premises so exceptional? In them, the imaginary Mrs. Smith becomes a grandmother at 36.

Have we not all a strongly analogous case before our eyes? Did not the illustrious lady, who now so happily fills the throne of these realms, become a grandmother at forty? Her eldest granddaughter is now in her twelfth year; and if it should please God to make Her Majesty's reign as exceptionally long as it has been exceptionally prosperous, she may be blessed by seeing her descendants to the fifth generation before she is called upon to celebrate her hundredth birthday.

I hope I may be pardoned for this allusion; but feeling that in this case the Queen's name is a tower of strength, I have ventured to use it in my endeavour to demolish the popular belief that a large number of descendants furnishes irresistible evidence of the abnormal age of those from whom they are descended.

Of the various kinds of evidence brought forward in proof of the great age of an alleged Centenarian, that which is founded on the supposed recollections of the old person, is at once the most fallible, unsatisfactory, and difficult to deal with—more especially in those instances where .these supposed recollections are brought forward in perfect good faith, and without any intention of deceiving, either on the part of the Centenarian or his friends. I believe the most conscientious self-examiner, when he comes to consider carefully what he believes to be his 'earliest recollection,' would find it very difficult to decide whether he really *recollected* such event, or having heard it much talked of in his youth, did not actually recollect it, but had it impressed upon his memory by what he had heard others say of it.

I was born in November 1803. Mr. Pitt died in January 1806, and his public funeral in Westminster Abbey took place on February 22, 1806. I was consequently only two years and three months old, when I was taken to see the procession ; and though I certainly have not the slightest recollection of any part of the ceremony, I have long believed that I recollect a slight personal incident on that day—namely, my father in his uniform as a volunteer bringing me a bag of buns! an incident well calculated to impress itself upon a child's memory. But do I recollect it ? I feel pretty sure that I do ; but my reason leads me to doubt whether it is not the recollection of what I heard, mixed up as it is with two or three other little matters connected with that ceremony, which must have been frequently talked of before me when I was a mere child.

The subject being one of great interest, I took advantage of a doubt suggested in 'Notes and Queries' (4th S. viii. 425), by the Rev. W. Sparrow Simpson, whether Dr. Johnson, being an infant of thirty months old, when he was touched for the evil by Queen Anne, would have even a 'confused' recollection of such an event—to invite (at p. 436) replies to the following queries :—

1. At how early an age is the mind of a child capable of being so impressed by any scene or event as to retain the memory of it till late in life?

2. Are there any means of distinguishing between the *bonâ fide* recollections of an individual of advanced age, and what such individual believes he recollects, but has in reality only heard talked of in his infancy by his relatives and friends?

But one answer, I think, was ever given to the second of these inquiries. It was signed F. C. H., and came from my late venerable friend, the Rev. Dr. Husenbeth, of Cossey, and was as follows :—

'As few things afford me greater pleasure than going back to early reminiscences, I am anxious to contribute my small share towards replying to the inquiries of T. He asks at how early an age is the mind capable of being so impressed as to retain the memory of any scene or event till late in after life? My own recollections probably are as clear and go as far back as those of any man approaching fourscore. I recollect dis-

tinctly places, persons, and occurrences, which I knew
when I was just turned of three years of age, and I
doubt if any one can remember anything that he witnessed at a much earlier age. T. inquires secondly, if
there are any means of distinguishing between *bonâ fide*
recollections of an aged person, and what he believes he
remembers, but has only heard talked of in his infancy.
I can answer only for myself; and I should say
that an old man is not likely to be mistaken as to
things that he himself saw, though he might be in
occurrences which he only heard spoken of. Thus I
have been often told of being for some time at the
seaside when only two years old, and asked if I did not
remember the name of my nurse, and playing with
dapstones on the shore ; but I always denied any actual
remembrance of these things, and never confounded
them with others which I remember perfectly as having
occurred soon after I was three years of age. Not long
ago I paid a visit to the house where I lived when three
years old, and had never once seen since, and I perfectly
astonished the present respectable occupant by recognising several parts of the premises which remain un-
changed, by saying, before I came to them, that certain
things ought to be here or there, by pointing out the
very spot where I saw a man kill a hedgehog, and
describing the melancholy march of poor French
prisoners along the road, with the baggage-waggons, on
which sat the wounded and women and children of the
soldiers, which took place in 1799. Others may remember occurrences when they were younger than

three, but I have never met with any whose recollection did actually reach farther back.'

But my first inquiry elicited some curious particulars, which, coming as they did, from correspondents whose good faith is above suspicion, may very fitly be recorded in this place.

The first of these (*ibid.* p. 483) refers rather to Mr. Simpson's observations than to my inquiry :—

' Mr. Simpson thinks that " an infant of 30 months old would surely not have even a confused recollection of such a visit." Will he permit me to inform him, as a case in point, that I have, not a confused, but a very distinct recollection of the rejoicings celebrated in my native village on the occasion of an important political event, at which time I was 26 months old ? Knowing this, I am unable to agree with him, but I readily admit that memory seems to commence with some persons at an earlier date than with others. Mr. Simpson adds, that Dr. Johnson " might certainly retain some recollection " of an event that happened when hé was four years and six months old. I should think that he might. I was less than that when I first saw Queen Mary's bedchamber at Holyrood, and I found my mental picture perfectly correct when I visited it sixteen years afterwards.

' I have a most distinct recollection not only of my nurse, who left us early, but of one room and particularly one cupboard and drawers, which I never saw or heard of after I was two years old, when we left the house. As neither the nurse nor the cupboard were

remarkable enough to be the subjects of after conversation, I am quite certain that my memory alone recalls
the objects seen.   ESTE.'

' My father, who was parochial clergyman at Dunino,
Fifeshire, had a new church built for him during my
early childhood. Building operations commenced when
I was about eight months old, and terminated before I
had reached eighteen months. I distinctly remember
having witnessed several men carrying the pulpit into the
church, a stone-hewer sculpturing a portion of the spire,
and the church bell hanging on a timber erection in the
churchyard. I might have been seventeen months old
when I witnessed these occurrences, but certainly not
older. During boyhood and in early youth my memory
was exceedingly imperfect, and still I am apt to forget
names just at the time when a recollection of them is
required most.   CHARLES ROGERS.'

' Snowdoun Villa, Lewisham.'

' As a sample of early recollections, here is a personal
one of my own :—I was born the latter end of June,
1834, and remember being taken to see the Queen
when Princess Victoria, at Kensington Palace. As
William IV. did not die till June, 1837, I cannot have
been three years of age. I went to the palace with my
mother and grandmother, and most distinctly remember
the Princess taking me into another room from that
where she received us, alone with her, and there giving
me an enamel ornament. So perfectly was this im-

pressed upon me that when I was a child the word
"princess" always meant to me a fair girl with curls,
dressed in white, holding up an ornament in her hand.
This is a perfectly tangible recollection. I remember it
as if yesterday. Other things I believe I remember
about the same time, but this is positive, and I do not
doubt T.'s two-year-old souvenir.　　　　E. J. C.'

'Though, like the Irishman, I was "by at the time," I
have no reminiscence of my birth-day, June 20, 1777;
the church registry is its only surviving evidence; my
transference, nine months later, to my grandfather's
residence in Worcester, is alike beyond my own, or any
other, authentication than long-extinguished hearsays.
This alone I can state on my positive and independent
remembrance:—My godfather, Sir Watkin Lewes—a
name even at this day not divested of its civic celebrity
—had long been the intimate friend of our family. My
"earliest recollection" of him is, that while he was our
guest a grand supper party was assembled, whereat I
was brought downstairs in my nurse's arms, and so
paraded up to my godfather at the upper end of the
table. The impression on my infant mind at the
sudden opening of the door, the lights, the company,
the long set-out, never through my protracted life has
left me. When at a later date I reverted to this
supper-scene, my father (he died in 1815) exclaimed, "I
remember it well; but, good God, you were then a
mere baby"

'EDMUND LENTHALL SWIFT.'

' I have a most distinct remembrance of carrying a small cat in my skirts from a farm-house to my own home, a distance of a mile and a half, when I was exactly two and a half years old. A prior remembrance is of some ferrets in a tub, and of my being told they were "blood-suckers." My grandfather always said that he could remember the birth of his brother, who was a year and a half younger.

' Thos. Ratcliffe.'

To these may be added a communication from Mr. Presley of Cheltenham, which appeared in ' Notes and Queries' of July 20, 1872.

'Early Recollections (4th S. viii.; ix. *passim*). A noteworthy instance is given in the very interesting "Life of Thomas Cooper," written by himself, lately published. He says :—

" I was born at Leicester on March 20, 1805 ; but my father was a wanderer by habit, if not by nature ; and so I was removed to Exeter when I was little more than twelve months old. I fell into the Leate, a small tributary of the Exe, over which there was a little wooden bridge that led to my father's dyehouse, on the day that I was two years old,—and, as my mother always said, at the very hour I was born two years before. After being borne down the stream a considerable way, I was taken out and supposed to be dead, but was restored by medical skill. It may seem strange to some who read this—but I remember, most distinctly and clearly, being led by the hand of my father, over St. Thomas's Bridge, on the afternoon of that day. He bought me some gingerbread from one of the stalls on the bridge ; and some of the neighbours who knew me came and chucked me under the chin, and said, ' How did you like it ?— How did you fall in ?—Where have you been to ?' The circumstances are as vivid to my mind as if they only occurred yesterday."

'To this I may add that my own memory carries me back at least to the day of her present Majesty's coronation, June 28, 1838, at which date I was 1 .day less than 2 years and 9 months old. I perfectly remember being carried by my grandfather through the streets of Bath to witness the illuminations, and also what some of the particular illuminations represented.

'JAMES T. PRESLEY.'
'Cheltenham Library.'

Of the good faith of the writers of these statements there can exist no doubt, and the statements themselves are so far valuable as illustrating a curious psychological question ; but that a confusion does exist in the minds of aged people as to what they recollect, and what they have heard is unquestionable ; and it is with the view of guarding against possible mistakes of this kind, in cases where there cannot be a doubt of the truthfulness and integrity of supposed Centenarians and their friends, that they are here reproduced.

How liable persons of the highest character, when of advanced age, are to confuse what they have heard with what they think they recollect, is curiously illus-

trated by a case with which I have been kindly furnished
by Lord Verulam.

' Dear Thoms,—With reference to what you said to
me to-day, I well recollect an old aunt of mine, who
lived to the age of 92, often mentioning circumstances
relating to the Rebellion of '45, as if she recollected
them herself. And when reminded that this was before
her birth, the answer was, ' But these things were the
great subject of conversation when I was young.'

' Yours truly,

'VERULAM.'

The remarks which we have just made apply more
particularly to those cases in which, whatever the real
facts may be, the truth and probity of those who bring
them forward are above suspicion.

Fortunately for the cause of truth, the pretended
recollections of pseudo-Centenarians are about the worst
witnesses they can bring into court. They generally
break down under cross-examination. ·

Centenarians of this class, like the lady in Hamlet,
' do protest too much'; and their pretended reminiscences furnish the most effective materials for exposing
how absolutely false are the very statements by which
they seek to establish the truth of their impudent pretensions.

The cases of old Geeran, Lahrbusch, and others, which
will be found fully detailed in the later part of this
volume, will establish this ; and I may now conclude

this part of my work with a few observations on the value of the evidence of those who profess to know that very old people are as old as they claim to be.

Old Smith is a hundred, and old Jones is sure of it, and he has known him all his life. Then old Jones is also a hundred. Prove that, and you go near to prove Smith's being of the age claimed. But when old Jones is questioned, he turns out to be 'hard upon eighty'—but he knows 'old Smith is a hundred, for he was a grown man when I was a boy.' Then comes the question of what age was the boy Jones, when Smith was a grown man; and it will generally be found he was about ten when Smith of twenty was a 'grown man,' and so old Smith proves to be 'hard upon ninety,' instead of a hundred; as had been stated, and believed because old Jones knew he was.

I was present lately at a conversation, in which the utter worthlessness of the recollection of one person as to the age of another was curiously shown. A question having arisen as to the age of a well-known dignitary, a gentleman said, 'From the time I have known him, and his age when I first knew him, he must be upwards of 80.' Adding, that he had known him nearly 40 years; that he was nearly 50 when he first became acquainted with him, which would have made him nearly 90. The question of dates was then gone into; and it appeared that the gentleman whose age was disputed was not nearly 90, was not upwards of 80, but was 73!

I have not made any reference to the cases of extreme Longevity which appear in the valuable Reports published from time to time by the Registrar-General of Births, Deaths and Marriages; and, for this simple reason, that the exceptional ages there recorded are, as a rule, unsupported by a single particle of evidence. The District Registrar is called upon to receive, and has no authority to investigate, even if he had the means and leisure, the truth of the statements which he is called upon and bound to record.

A curious illustration of this has occurred very recently. In ' The Times'' Obituary of 20th September appeared the following announcement: ' On the 12th Sept. at S. James's Road, Holloway, William Highgason, aged 107 years.' The age was so exceptional that I was sure there must be some error in the statement; and being unable to go to Holloway, I got a friend to make some inquiries for me; the result of which showed that the daughter, acting on the information of her brother-in-law, gave the Registrar the age of her father as 107. But the widow, who survives him, does not believe that he had reached that age; for, according to the age given to her at their marriage, he would be 98 at his death; and if he purposely understated his age at that time by three years, (as she *thinks* he did) he would have been 101 when he died. He was born either at Enfield Chase, or Shenley Hill, Herts; most probably the former, as the Registers at the latter place have been searched for his baptismal certificate without success.

From the 33rd Report of the Registrar-General lately issued, it appears (p. xii.) that during the year 1870 the deaths of 18 males and 63 females whose ages were *stated to be* upwards of 100 years were registered—in all 81 persons, against 63 and 79 respectively in 1868 and 1869. The highest ages said to have been attained in 1870 were 108 by a male and 107 by a female. Of these 18 males, 6 are said to have been aged 100 ; 1 aged 101 ; 3 aged 102; 2 aged 103 ; 5 aged 104; and 1 aged 108. While of the females, 22 are said to have been aged 100; 13 aged 101 ; 11 aged 102 ; 6 aged 103 ; 8 aged 104; 2 aged 105; and 1 aged 107 ; and I am strongly inclined to believe that out of the whole of these 81 cases, there is scarcely one which would stand the test of investigation—scarcely one which would not prove to rest upon as little foundation as the 107 years of William Highgason.

And the Registrar-General, with great judgment and propriety, does not do, what the public too frequently and mistakingly credit him with doing, namely, declare, that 18 males and 63 females aged 100 and upwards died in England during the year 1870 ; but, that during such year the deaths of so·many males and so many females have been registered, ' *whose ages were stated* to be upwards of 100 years.'

Nay more, when the Department has the oppoi tunity of investigating any case of supposed exceptional Longevity, the result is published in ' The Weekly Return.' Thus the case of George Fletcher, stated to have died in 1855 at the age of 108, was shown in ' The Weekly Return ' of 17th February, 1855, from the investigation of Dr. Farr, to have been only 98 : while on the other hand, 'The Weekly Return' for July, 1870, furnishes clear and decisive evidence that Mr. Jacob William Luning was really upwards of 103 at the time of his death.[1]

In the Report of the Registrar-General prefixed to the Census of 1851, in which he points out the Results and Observations deducible from the various Tables, there is a very important passage on this subject which has never met with the attention it deserves. It is as follows:

'At the last Census (1851) 111 men and 208 women have been returned of ages ranging from 100 to 119 years; and to the scientific inquirer in the districts where these old people reside, an opportunity is afforded of investigating and setting at rest a problem of much greater interest, than some of the curious questions that engage the interest of learned societies. Two-thirds of the Centenarians are women. Several of them in England are natives of parishes in Ireland or Scotland, where no efficient system of registration exists; few of them reside in the parishes where they were born, and have been known from youth; many of the old people are paupers and probably illiterate;—so that it would no doubt be difficult to obtain the documentary evidence which can alone be accepted as conclusive proof of such extraordinary ages.'

It is much to be regretted that 'the scientific inquirers in the districts in which these old people resided' turned a deaf ear to the suggestion of the Registrar-General,

---

[1] For full particulars of both these cases, see the later portion of this volume.

and did not endeavour to investigate and set at rest, problems, which, despite of my own former dallyings with Primeval and Mediæval Antiquities, I admit to be of much greater interest — inquiries too which can be pursued with so much success by residents on the spot.

I confidently believe that if any such well organised inquiry had been undertaken in 1851, the public mind would by this time.have been impressed generally with such correct ideas on the subject of Human Longevity as would have rendered the appearance of the present work superfluous and uncalled for.

CHAPTER V.

HAVING shown in the preceding chapters the caution with which, not only all unsupported statements of abnormal Longevity are to be received, but that such caution should be extended to the various species of evidence brought forward in support of cases supposed to be clearly established, I propose to illustrate the value of such caution by giving the results of its application to a variety of cases of Longevity which I have investigated.

But before doing so, I will endeavour to remove from the paths of the honest inquirer after truth those stumbling-blocks, the 169 years of Henry Jenkins, the 152 years of Old Parr, and the seven score years of the Old Countess of Desmond. As Henry Jenkins is the greatest offender against probability and common sense, I will begin with him.

It is now several years since, upon my attention being accidentally called to the subject of abnormal Longevity, I came to the conclusion, strengthened and confirmed by subsequent inquiry and consideration, that there was not the slightest evidence in support of the statements so frequently and confidently repeated, that Henry Jenkins had lived to reach the incredible age of 169, Old Parr that of 152, and the celebrated old Countess of Desmond that of 140.

But it was not until a later period that I took any steps to investigate the two former cases ; and I believe a letter of inquiry, printed in ' Notes and Queries,' of May 21, 1870 (4th S. v. 487), in which I inserted the request ' Yorkshire papers please copy,' a request with which several of them very kindly complied, was the first doubt as to Jenkins' age that had ever been publicly avowed. The following is an extract from such letter :—

' I am now anxious, for a particular purpose, to make a similar appeal[1] to your readers and to Yorkshire antiquaries generally, for any evidence they may possess in confirmation of a single statement in the yet more marvellous story of Henry Jenkins ; and for obvious reasons I should like here to borrow a phrase now frequently added to announcements in newspapers— " Yorkshire papers, please copy."

' Jenkins *is said* (but not the slightest authority has ever been produced for the statement) to have been born in 1501. He died "a very old man," says the parish register, and was buried December 9, 1670.

' The earliest account of Jenkins appears to be that given by Miss Savile, which, though not dated, is be- lieved, on reasonable grounds, to have been written about 1662 or 1663. According to that account,

---

[1] To that which in the preceding year I had addressed to Salopian Antiquaries about Old Parr. See page 88.

Jenkins was at the time, "to the best of his remembrance, about 162 or 163."

'On April 15, 1667, when examined at Catterick, he is described as "Henry Jenkins, of Ellerton-upon-Swale, labourer, aged 157, or thereabouts."

'In Miss Savile's report he is described as having "sworn as a witness in a cause at York to 120 years—which the judge reproving him for, he said he was butler at that time to Lord Conyers."

'Sir R. Graham mentions that "Jenkins gave evidence to six score years in a cause between Mr. How and Mrs. Wastell, of Ellerton." Is anything known of this cause? in speaking of which Mr. Clarkson, in his "Richmondshire," tells a remarkable story of Mrs. Wastell's agent, on going to summon Jenkins, finding at Ellerton a son and grandson alive, both of whom were much more infirm in memory than Jenkins.

'What is Mr. Clarkson's authority for this? and when did this son and grandson die?

'Jenkins's wife died in 1668. Was she his first wife? When and where were they married? What was her age? Her death, and that of her husband, are said to be the only two entries in which the name of Jenkins occurs in the register of Bolton.

'I have also seen mention made of Jenkins's evidence in a cause in 1667 between the vicar of Catterick and John and Peter Mawbank. What is the authority, and where is there any record of such trial?

'Yorkshire antiquaries may be in possession of other facts in reference to the alleged longevity of Henry

Jenkins. If so, I trust, in the interest of historical truth, they may kindly bring them forward.

'I have little hope of being able to prove the age of this Yorkshire patriarch ; what I do hope to accomplish with respect to him will be greatly promoted by any fresh and trustworthy information about him.'

In that letter, as will be seen, I appealed not only to the readers of 'Notes and Queries,' but to Yorkshire antiquaries generally, for any evidence they might possess *in confirmation of a single statement*, in any one of the marvellous long stories told of Henry Jenkins ; and I indicated certain specific points, on which, if true, corroborative evidence might reasonably be expected.

With the exception of a reference to the 'Yorkshire Archæological and Topographical Journal,' in which the Rev. Canon Raine has provided for the first time *accurately*, what had before been very imperfectly given, one of the documents which is supposed to furnish evidence of Jenkins's longevity, I did not elicit a single communication.[1]

From this, it may fairly be concluded that all the evidence which exists as to the age of Jenkins is before us ; and what does it amount to? My readers will hardly be prepared for the assurance that the belief in

---

[1] I speak here of *evidence*—not of popular belief and gossip. Much of this will be found duly recorded in a carefully prepared volume, entitled, *Evidences of the great age of Henry Jenkins, with Notes, respecting Longevity and long-lived Persons.* Bell, Richmond, 8vo, 1859. This is obviously the production of a gentleman possessed of great local knowledge, and desirous of arriving at the truth, but unaccustomed to sift evidence.

the marvellous age of Henry Jenkins, which has so long
existed unchallenged, and so influenced speculations on
the duration of human life, rests upon no better evidence
than JENKINS'S OWN ASSERTION !

With the exception of the register of his burial,
which does not state how old he was at the time of his
decease, there exist but two documents in which the
age of Jenkins is mentioned, and in both of these the
age stated is what he *himself declared.*

The case of Henry Jenkins forms no exception to the
rule which governs the majority of cases of *extreme*
Longevity—namely, that it is entirely based upon
the unsupported testimony of the supposed Centenarian.

If Jenkins were at the time of his death in 1670, 169
years of age, he must have been born in·1501 ; and yet
there is not the slightest trace of the existence of
the man who had been 'butler to Lord Conyers,' and
' was often at the assizes at York,' until in his 162nd or
163rd year, he related to Miss Savile the absurd story
of himself, on which the popular belief of his great age
is mainly founded; and this is the more remarkable,
since more than twenty-five years previously—namely
in 1635—public attention had been called to the subject
of Longevity, by Taylor the Water Poet's account of
Old Parr.

Although the account which Miss Savile has left us,
and which will be found reprinted from the ' Philosophical Transactions' at length, and I hope with
greater accuracy than usual, in the Appendix, is not

dated, it is clear that the interview between the lady and Jenkins took place in 1662 or 1663.

In this account, she tells us that when she came first to live at Bolton, 'they told me there lived in that parish a man near 150 years old; that he had sworn as *a witness in a cause at York to* 120 *years*, and that the judge reproving him, he said he was butler in that time to Lord Conyers, and they told me that it was reported his name was found in some old register of the Lord Conyers' menial servants, but truly it was never in my thoughts to inquire of my Lord Darcy whether this last particular was true or no ; for I believed little of the story for a great many years ; till one day, being in my sister's kitchen, Henry Jenkins coming in to beg an alms,' after exhorting him, as he was an old man who must soon give an account to God of all he did or said, she desired him to tell her very truly how old he was ; 'on which he paused a little, and then said, to the best of his remembrance he was about 162 or 163.' He then told her he remembered Henry VIII. and the battle of Flodden ; that the King was not there, as he was in France, and that the Earl of Surrey commanded. Being asked how old he was then, he said, ' I believe I might be between 10 and 12, for,' says he, ' I was sent to North Allerton with a horseload of arrows, but they sent a bigger boy from thence to the army with them.' Finding Jenkins's story about Flodden, when it was fought, &c., confirmed by old chronicles, and that if he was 10 years old when it was fought, 'he must be 162 years or 163, when I examined him,' the lady remarked,

'so that I don't know what to answer to the consistency
of these things, for Henry Jenkins was a poor man,
and could neither write nor read. There were also four
or five in the same parish that were reputed all of them
to be 100 old or within two or three years of it, and they
all said he was an *elderly* man, ever since they knew him,
for he was born in another parish, and before any
registers were in churches, as it is said ; he told me then,
too, that he was butler to the Lord Conyers, and remembered the abbot of Fountains Abbey very well, who
used to drink a glass with his lord heartily, and that the
dissolution of the monasteries he said he well remembered.'

Dr. Tancred Robinson, who sent this letter in 1696 to
the Royal Society, accompanied by some remarks, tells
us, that ' Jenkins in the last century of his life was a
fisherman, and used to wade in the streams ; his diet
was coarse and sower ; he had *sworn in Chancery and
other Courts to above* 140 *years' memory* ; and was often
at assizes at York, whither he generally went a-foot,
and I have heard some of the country gentlemen affirm
that he frequently swam in the rivers when he was past
the age of 100 years.'

From this it will be seen that in 1662 or 1663,
Jenkins asserted his age to be 162 or 163, that he was
' between 10 and 12 when Flodden was fought, which
corresponded with his former statements ; and though
Miss Savile does not say so, it may fairly be inferred
that Jenkins, on that account at least, computed his
age from that event.

The next document is of four or five years' later date, being Jenkins's deposition at Catterick, in April 1667, when he was called as a witness in a tithe cause, between the Rev. Charles Anthony, Vicar of Catterick, and Calvert Smithson, one of his parishioners. The important parts of this document will also be found in the Appendix.

Jenkins was the eighth witness, and he is described in the official record, as 'Henry Jenkins of Ellerton-upon-Swale, in the County of Yorke, Labourer ;' and in his deposition he deposes, 'that to this deponent's knowledge, all the particulars mentioned . . . . were payed in kinde . . . . above three score yeares ;' a fact to which any very old man, as Jenkins undoubtedly was, might very reasonably depose.

But the reader will here be struck by a very strange discrepancy in the old man's account of himself. Four or five years before he had told Miss Savile he was 162 or 163 ; but reversing the ordinary course of time, instead of being three, four, or five years older, he now declares himself to be actually five or six younger, that is to say, only 157.

But the longest life, be it what it may, must have an end, and that of Henry Jenkins came in December 1670 ; and the register of burials for Bolton-on-Swale for that year, under the date, December 9, contains the following brief, and no doubt truthful entry :—

'Henry Jenkins, a very aged and poore man, of Ellerton, buried.'

His wife had predeceased him only a very few years, having been buried on January 27, 1667–8. Canon Raine states that these are the only notices of the family of Jenkins, which the Bolton parish register contains.

When it is considered how widely spread, as has been shown, was the popular belief in Jenkins's extraordinary age, one is a little surprised to find that his age is not recorded in the register, unusual as it may have been to record the ages of the parties buried ; and that he should simply be described as 'very aged;' and the surprise will be increased when we find, as we do, from the pamphlet to which I have already referred (p. 6 or 7), that the register, which is carefully kept, is in the handwriting of the vicar of Catterick,—the very Charles Anthony in whose favour Jenkins had appeared as a witness in the vicar's successful suit against Smithson, when Jenkins declared himself to be 157.

Canon Raine describes Anthony as a 'strict, exact man, and evidently a very careful parish priest.'

Surely this is a pretty strong indication that in the opinion of this exact and careful man, who knew Jenkins well, although he was a 'very aged man,' how aged he was was very doubtful.

But it may be asked, and with some show of reason, If this is all the evidence which can be produced in support of Jenkins's 169 years, how comes it that the story of his great age has been so widely spread and remained so long unchallenged? This is a question difficult to answer, and which probably cannot be better met, than in the observation of Dr. Johnson, who in his 'Journey to the Hebrides,' 8vo, 1791, p. 192, among other pithy remarks on the subject of Longevity, says, 'Instances of long life are often related, which those who hear them are more willing to credit than to examine.'

It is to be borne in mind, moreover, that Jenkins was gifted with a most remarkable memory. We have notices of this not only in the cause of Anthony v. Smithson, where he modestly confined himself to deposing 'to above three score years;' but in Miss Savile's account of him she tells us 'he had sworn *as a witness in a cause at York to* 120 *years*, which the judge reproving him for' (as very well he might), 'he said he was butler at that time to Lord Conyers;' while Dr. Robinson caps Miss Savile by gravely asserting 'he had *sworn in Chancery and other Courts to above* 140 *years' memory, and was often at the assizes in York*, whither he generally went a-foot;' so that without impugning the truthfulness of Henry Jenkins, or insinuating that his memory was a convenient one, and one which was a source of emolument to him, it is a fair inference that those who benefited by his testimony would, as a matter of course, be ready to believe and insist upon his being every year of the many to which he laid claim.

Nay, have we not in our own days, seen a man of literary and scientific attainments not only believe all this, but even go beyond it. We have already alluded to his avowed doubts whether in 'Cases of Longevity,' the testimony of contemporaries is not of more value than a mere register of births and deaths; and in the same paper ('Good Words,' July 1865, p. 493), as if gifted with a prophetic foresight of my incredulity with regard to the case of Henry Jenkins, he writes:—

'If, however, sceptics must have documentary evidence of a circumstance which was patent to the whole country side, we have the best of all such proof in the fact that THE REGISTERS OF THE COURT OF CHANCERY PROVE THAT HE GAVE EVIDENCE ONE HUNDRED AND FORTY YEARS BEFORE HIS DEATH.'

Perhaps it is scarcely necessary that I should add that I am responsible for the small capitals in this extract; and that I have not asked my kind friend Sir Thomas D. Hardy, to have the Chancery Registers searched for the remarkable entry referred to.

But, for the credence which has so long been given to this grossly incredible fable (and grossly incredible it really is when carefully looked into), I fear the Royal Society is somewhat responsible.

When that learned body gave insertion in the 'Philosophical Transactions' (vol. xvii., pp. 266–8, No. 221. 1696) to Dr. Tancred Robinson's account of Jenkins, accompanied by a note, which simply expressed a wish that 'particular inquiries should be made and answered concerning the temperance of this man's body, manner of living, and all other circumstances, which might furnish useful information to those who are curious about Longevity,' but without adding one word of doubt as to his reputed age of 169, it affixed the seal of its great authority to the truth of the story.[1]

But this gentleman does not stand alone in maintaining with more zeal than judgment, the 169 years of Henry Jenkins.

In 1743, the good people of Yorkshire, proud as they justly are of everything connected with their county, and as jealous of the fame and age of Jenkins as the Salopians are of Old Parr and his 152 years ; raised a subscription and erected in Bolton Churchyard an obelisk to the memory of Jenkins, which records his name, assumed age, etc., and so ' like a tall bully, lifts its head and lies ;' and placed in the church a black marble tablet, on which is engraved the following inscription, alike grandiloquent and unfounded, a striking contrast to the simple truthful record of the parish register.

---

[1] I quote in support of this the following passages from the ' Life of Jenkins' already referred to—

' Jenkins's fame in his own neighbourhood would be kept up by the paper read before the Royal Society ; that society was then popular and fashionable ; and Dr. Robinson, a distinguished Naturalist and Court Physician, p. 14.—Again, a little further on, ' The publication of Miss Savile's Letter, and the erection of the monument in Bolton Church, would be a sort of double test and ·challenge to all who might be inclined to dispute the matter.'—*Ibid.*

While it is obvious from the cautious manner in which the Registrar-General speaks of these ' extraordinary instances' (alluding to Jenkins and Parr) in his Report on the Census of 1851, and from the care with which he quotes the Phil. Trans. as his authority, that his judgment was influenced by the manner in which the Royal Society had accepted their reputed ages.

BLUSH NOT MARBLE,
TO RESCUE FROM OBLIVION,
THE MEMORY OF
HENRY JENKINS,
A PERSON OBSCURE IN LIFE,
BUT OF A LIFE TRULY MEMORABLE;
FOR
HE WAS ENRICHED
WITH THE GOODS OF NATURE,
IF NOT OF FORTUNE;
AND HAPPY
IN THE DURATION,
IF NOT VARIETY
OF HIS ENJOYMENTS;
AND
THO' THE PARTIAL WORLD
DESPISED AND DISREGARDED
HIS LONE AND HUMBLE STATE,
THE EQUAL EYE OF PROVIDENCE
BEHELD AND BLESSED IT,
WITH A PATRIARCH'S HEALTH, AND LENGTH OF DAYS,
TO TEACH MISTAKEN MAN,
THESE BLESSINGS ARE ENTAILED ON TEMPERANCE,
A LIFE OF LABOUR, AND A MIND AT EASE.
HE LIVED TO THE AMAZING AGE OF
169.
WAS INTERRED HERE DECEMBER 6TH,
1670;
AND HAD THIS JUSTICE DONE TO HIS MEMORY,
1743.

I have seen this monument referred to as incontestable evidence that Jenkins was really 169; for if the statement of his great age had not been true, would the shrewd men of Yorkshire have subscribed to put up such a monument? I have unfortunately mislaid the reference. I thought it was in Whitaker's ' Richmondshire,' but it is not. In searching there for it, however, I find a passage in which the author gives his opinion on the story of Jenkins, which may well be inserted here as a proof of the error of judgment into which

an earnest painstaking inquirer after truth may be led
by unconscious prejudice, an innate love of the marvellous, or inability to weigh and examine evidence :—

'After all, I have thought it worth while to consider,
whether, in the absence of a baptismal register, the
astonishing longevity of Jenkins is really ascertained
by collateral evidence. The curious and perfectly
credible evidence of the man with respect to the
messages which he bore between the old Lord Conyers
and Marmaduke, abbot of Fountains, proves only that
he was of age to bear such messages a few years before
the dissolution, and might therefore have left him
twenty years younger ; but the distinct and particular
narrative of his having been entrusted with the carriage
of arrows before the field of Flodden, leaves no room
for doubt. A boy of less than thirteen would not have
been sent on such an errand. In the reign of Charles II.,
and for many years before, Jenkins alone survived to
tell in the ears of a generation wholly indifferent to an
event so long past and gone, the universal grief and
consternation which prevailed in Richmondshire on the
dissolution of the religious houses.

'Excepting that his memory was retentive, nothing
else has been recorded with respect to the understanding of this wonderful man ; but for obvious reasons,
instances of very great Longevity scarcely ever occur
but in a rank of life, where a few leading and striking
facts alone are preserved, but where there is too little
curiosity or power of reflection to mark the progressive
modes and changes of human life.' [1]

[1] ' Whitaker's Richmondshire,' 1823, vol. ii. pp. 39, 40.

But I may be told I have omitted to notice one of the well-known cases of Jenkins's having given evidence at York—namely, in the cause of Howe *v.* Wastell, and the extraordinary circumstances connected with it, so distinctly stated by Mr. Clarkson in his ' History and Antiquities of Richmond ' pp. 396-7, that there can be no doubt Mr. Clarkson was satisfied of their truth. The statement is as follows :—

' Previous to Jenkins's going to York, when the agent of Mrs. Wastell went to him, to find out what account he could give about the matter in dispute, he saw an old man sitting at the door, to whom he told his business. The old man said, ' He could remember nothing about it, but that he would find his father in the house, who perhaps could satisfy him.' When he went in he saw another old man sitting over the fire, bowed down with years, to whom he repeated his former questions. With some difficulty he made him understand what he had said; and after a little time got the following answer. which surprised him very much. ' That he knew nothing about it, but that if he would go into the yard, he would meet with his father, who perhaps could tell him.' The agent upon this thought that he had met with a race of Antediluvians. However into the yard he went, and to his no small astonishment found a venerable old man, with a long beard and a broad leathern belt about him, chopping sticks. To this man he again told his business. and received such information as in the end recovered the royalty in dispute.'

On this I need only remark that Mr. Clarkson does

not adduce a single authority to show that such a suit
was ever instituted. His statement is founded probably
on what Sir Henry Graham has said, 'that Jenkins gave
evidence to six score years in a cause between Mr. Howe
and Mrs. Wastell, when Sir Richard Graham was sheriff.'
The answer to this is very simple. Sir Richard was
not sheriff till 1680, ten years after Jenkins's death,
and Mrs. Wastell's husband did not die till the year
after Jenkins, that is, 1671. But this is not all, Jenkins
is never said to have left any family, and the Register
of Bolton contains no mention of any such.

The fact is that the story of Jenkins's son and grandson is only a Yorkshire version of the story as old or
older than Jenkins himself—namely of the very old
man who was seen crying because his father had beaten
him for throwing stones at his grandfather!

But there is something yet more startling to be laid
before the reader. In the year 1865 there was living in
Edinburgh an old gentleman who had seen a man who
had seen a man who had seen Jenkins! and this gentleman furnished the following extraordinary statement
to the ' Edinburgh Courant.'

'THREE CENTURIES AND A HALF AGO.—' I have
seen a man who conversed with *a man who fought* at
Flodden Field,' may be said by a venerable octogenarian
gentleman to whom we are indebted for the following
most interesting memorandum:—The writer of this, when
an infant, saw Peter Garden, who died at the age of 126.
When 12 years old, on a journey to London about the
year 1670, in the capacity of page in the family of Garden

of Troup, he became acquainted with the venerable Henry Jenkins, and heard him give evidence in a court of justice at York, that he "perfectly remembered being employed, when a boy, in carrying arrows up the hill at the battle of Flodden."

<pre>
        ' It was fought in   .   .   .   A.D. 1513
   Add Henry Jenkins's age .   .   . 169
          Less   11
                                    ───
                                          158
   Peter Garden 126
      Less his age when at York.   . 12
                                    ───
                                          114
   The writer of this in 1865, aged .   .   .   80
                                         ─────────
                                         A.D. 1865.'
</pre>

This was just the paragraph to go the round of all the papers. It appeared in 'Notes and Queries' of October 21, 1865 (3rd S. viii. p. 329), where attention was called to the fact that in this new and improved version of the story of Jenkins' appearance at York he is described as 'a *man* who had *fought* at Flodden.' His own improbable statement was, that he remembered Flodden Field, when 'he was sent to North Allerton with a horseload of arrows, but they sent a *bigger boy* from thence to the army with them.'

We are next told that the link between Jenkins and the octogenarian is 'Peter Garden,' who died at the age of 126 (?) 'and on a journey to London about 1670,' 'became acquainted with Jenkins,' and 'heard him give evidence in a court of justice at York, that he perfectly remembered being employed when a boy in *carrying arrows up the hill at the battle of Flodden.*' It was very lucky Peter Garden was at York in 1670, for in that very year Jenkins died; and though it is said in the accounts of him that he was 'often at the assizes at York,' the only recorded evidence of his which is in existence, was given in a case at Catterick in 1667, and in that evidence there is not a word about Flodden.

But how the Octogenarian, who only saw Peter Garden, knows all he tells us about that venerable person; or how it happened that he saw him at all is a matter of great interest. For though there is no evidence that Peter Garden *was* 126, or as to where or when he was born, we learn from Easton's ' Human Longevity' that he died in 1775, just ninety years before 1865. How a gentleman who was only an octogenarian in that year could have seen Peter Garden, who *died before any octogenarian living in* 1865 *was born*; and how the boy who was only sent to North Allerton with the arrows got to Flodden (several score miles distant), *carried the arrows up the hill,* and became *a man who fought there,* are only some of the many contradictions and absurdities in this strange story, which it will be for the correspondent of the ' Edinburgh Courant' to explain.

Absurd and exaggerated as all this is, it is only of a piece with the whole story of Henry Jenkins and the 169 years with which he has hitherto been credited without a particle of evidence in support of them.

I hope the time is not far distant when the reputed age of Henry Jenkins will no longer interfere with scientific inquiry into the average duration of Human Life.

CHAPTER VI.

IF in doubting the 169 years of Henry Jenkins, I have
been guilty of an act of daring scepticism ; what can be
said in extenuation of my still greater audacity in doubting the 152 years of Thomas Parr?—proved as they are
supposed to be by his being presented to the King by
Lord Arundel, by his life by Taylor the Poet, and by
his monument in Westminster Abbey ; and accredited,
as they have been supposed to be, by the testimony of
Harvey.

I have only one excuse to offer ; but that is a valid
one.   I am right.

There is no doubt that Thomas Parr was a very old
man, an exceptionally old man ; probably a hundred ;
possibly a year or two more.   But I do not believe that
there exists a single particle of evidence in support of
the monstrous fable which has been so long and so
readily believed.

Of Old Parr, the story runs, that he was born at
Winnington in the parish of Alberbury, in the year
1483.   In the year 1635, the Earl of Arundel, visiting
his estates in Shropshire, 'the report of this aged man'
(he was supposed to be then 152), 'was certified to him.'

He saw him, and eventually sent him in a litter to London, where, about the end of September, he was presented to the King. On November 15 he died, and on the following day, Harvey made the celebrated post-mortem examination of him, which is so often referred to ; and then the ' old old, very old man,' was buried in the south transept of Westminster Abbey, where his gravestone records what no doubt those who placed it there, honestly believed, that he had ' lived in the reigns of ten kings.' This statement has no doubt contributed to maintain the popular belief in his 152 years.

I have printed, in the Appendix to this volume, Taylor's Life of the old man (more accurately I trust than in any former reprint) ; Harvey's ' Anatomical Examination' of him ; and one or two other *pièces justificatives*, to which I refer in this notice, in order that those who may differ from me in my conclusions, may have the means of testing my authorities, and of correcting any error into which I may have inadvertently fallen.

Parr's case differs from that of Jenkins in so far, that while the latter rests mainly on the old man's own authority, Taylor, in his metrical life of Parr, presents us with such a multitude of reputed dates and facts, that it seems almost impossible, if they are correct, but that confirmatory evidence of some should exist. What authority beyond ' common report' Taylor had for these statements does not appear. He might possibly have learned some from Parr's ' daughter-in-law, named Lucy,' who accompanied him to London ; some possibly from the document to which he refers at the close of his poem :—

And gentlemen o' the country did relate
T' our gracious king, by their certificate,
His age, and how Time with gray hairs hath crown'd him.

When I first began to examine the case of Thomas Parr, it seemed to me highly probable, looking to the numerous dates and facts so distinctly recorded with respect to the chief events of his life, that some documentary trace of his existence might be found, from which it might be possible to gather trustworthy information respecting him and his age. In my anxiety to discover some of these, I am afraid I not only tired the patience of my friends, but that of many other gentlemen, with whom I had not the advantage of being acquainted, and to whom I had only ventured to apply on the broad ground of the general interest of the subject, and its bearing upon an important scientific truth.

My most kind old friend, the Rev. John Webb—and all who knew that ripe scholar and accomplished antiquary will know with what zeal he would enter on such an inquiry—made several searches among the diocesan records at Hereford, but to his great regret and disappointment, without eliciting one scrap of information.

The Vicar of Alberbury, after examining the registers, assured me, what I had already been told, that no mention of Parr, is to be found in them ; a fact lately confirmed by the Rev. John Pickford, who examined not only the registers, but various books and papers in the church chest, without success.

Foiled in every endeavour to obtain the information of which I was in search, by means of private applica- 'tion, I determined to make a public appeal to Old Parr's fellow countrymen ; in the hope that such a step might be attended with better results.

The letter, from which the following is an extract ap- peared in ' Notes and Queries' of June 26, 1869 (4ᵗʰ S. vii. p. 569) ; and I desire here to acknowledge the great courtesy which I received at the hands of the local press, for I believe that without any exception, my appeal was transferred to the columns of the leading journals of Shropshire and the surounding counties :—

' The fullest account we have of Thomas Parr is con- tained in the metrical life of him by Taylor the Water Poet, published in 1635. · Upon what authority Taylor founded his very definite statements as to the events of Parr's life, and the dates at which they occurred, does not appear. Probably the same common report, to which Hervey referred, or some broadside circulated and believed at the time. But these statements are, under the circumstances of Parr's rank and condition of life, exceptionally remarkable for precision and minute- ness, as may be seen by the following abstract :—

' 1483 is set down as the year of the birth of Thomas Parr, the son of John Parr, of Wilmington.

' In 1500, Parr being then 17 years of age, went into service, in which service he continued for eighteen years ; when,

'In 1518, being then 35, Parr returned home, as may be inferred upon the death of his father, since we are told—

> . . . his sire's decease,
> Left him four years' possession of a lease.

'In 1522, he being then 39, Parr received a new lease from Mr. Lewis Porter.

'In 1543, Parr, being then 60, got a further lease from Mr. John Porter, son of Mr. Lewis Porter.

'1563, Parr being then 80, married his first wife, Jane Taylor, a daughter of John Taylor, by whom he had two children—a boy, John, who died when only 10 weeks old ; and a daughter, Joan, who lived only 3 weeks.

'In 1564, Parr, being then 81, obtained a fresh lease from Mr. Hugh Porter, the son of Mr. John Porter.

'In 1585, Parr, being 102 years old, obtained from—

> . . . John, Hugh's son,
> A lease for 's life, these fifty years outrun.

'In 1588, Parr, being then 105, did penance in a white sheet in Alberbury church for having had a bastard child by Katherine Milton.

'In 1595, Parr, being then 112, buried his first wife, Jane, to whom he had been married for thirty-two years.

'In 1605, Parr, who was then 122, having been a widower for ten years, married his second wife, Jane, daughter of John Floyd ("corruptly Flood," says his biographer), of Gillsells, in Montgomery, and widow of Anthony Adda.

'On November 14, 1635, Parr died, having, as it is alleged, attained the remarkable age of 152 years, 9 months, and some odd days !

'Such is the incredible story told of the "old, old, very old man;" and I really hardly know which is the more to be wondered at—the exceptionally great age of 152 attributed to Parr; or the fact that for upwards of two centuries nobody has appeared to doubt its accuracy, or to have taken the slightest trouble to ascertain upon what evidence it is founded.

'I have personally, and with the assistance of several kind friends, made many endeavours to find any evidence which might throw light upon the age which Parr had actually attained; but all my efforts have hitherto proved fruitless.

'Although my endeavours to discover the slightest corroboration of any one of the facts relating to Old Parr, with the exception of that of his death in 1635, have utterly failed—to the strengthening of my entire disbelief in his alleged Longevity—it has occurred to me that an appeal to the readers of "Notes and Queries," and more especially to such antiquaries, men of letters, and clergymen as may be connected with Shropshire or interested in its history, might be productive of better results.   I therefore venture to make this public appeal for information of any kind calculated to throw light upon the real truth of the story of Thomas Parr—a story in its present form incredible in itself, unsupported by evidence, and inconsistent with all the known laws of physical science.

'Let me add, that I am not asking for references to ordinary books. I believe I am in possession of references to most, if not all, the printed authorities on the subject of Old Parr.'

My readers will scarcely believe it, but this public and widely circulated request for information was as little successful as my private applications had been ; and utterly failed in bringing out one single scrap of evidence.

It certainly did call forth *one* reply ; which deserves notice, if only as an amusing specimen of the 'snubbing' which any one who ventures to doubt a popular belief, may make up his mind to receive.

D. D. writing from the 'Abbey, Shrewsbury,' only a few miles from the scene where Parr passed his life, and where any existing records of it might be looked for, says, 'I was born in the parish of Alberbury, and am now an old man ; but can remember in my childhood how very much I then heard about Parr. His history had been handed down through many generations without the slightest attempt at exaggeration. I have been in the neighbourhood again, last week, and find the present inhabitants giving the same account as I heard an age ago ; but I never heard that Old Parr was such an idiot as to swallow pills to preserve his health with ; and if any one of the present generation who does swallow such stuff should live to be half the age of Old Parr, it will be a greater miracle than anything in Parr's history. Alberbury Church adjoins Luton Hall, the residence of Sir Baldwin Leighton, and, *I have no doubt*, the particulars of Parr's penance may be found in the church records, *to which I would refer all sceptics.'*

D. D.'s reference to the 'particulars of Parr's penance,' to be found in the records of Alberbury Church, is a great climax to his indignant protest against 'all sceptics ; ' and would be triumphant, only unfortunately these 'particulars,' which are the very things wanted, do not exist.

In the absence of a single scrap of information in support of any one of the minute particulars recorded of the ' old, old man,' it seems impossible to arrive at any other conclusion than that the particulars in question have no other foundation than idle gossip; unless in some cases the Water Poet is open to the charge which Sheridan once made against a political opponent, that ' he drew upon his imagination for his facts.'

But I may be told that Taylor certainly had authority for his statement :—

> That for law's satisfaction 'twas thought meet
> He should be purged by standing in a sheet ;
> Which, aged he one hundred and five years,
> In Alberbury's parish church did wear.

It is true. In the curious account, which 'Mr. Harrison, a painter of Norfolk,' gives of Parr, whom he saw during the two days the old man was staying with the Earl of Arundel at Wem, he tells us—' The King said to Old Parr, "You have lived longer than other men, what have you done more than other men ? " " I did penance when I was a hundred years old." The same he told me before he went to the King ! '

But the reader will remember that, like Jenkins's story of taking the arrows to the army at the time of the Battle of Flodden, this is Parr's own account; and we have nothing but his own word for an incident which, if true, is utterly inconsistent with other parts of his story. For if the old fellow's animality—(to use an expressive word which we owe to the late Mr. Henry Crabb Robinson) was so strongly developed as to lead to his public punishment for incontinency when he was a hundred and upwards, his first wife being then alive, and to his boasting of it half a century afterwards, it is scarcely conceivable, that

> A tedious time a bachelor he tarried,
> Full eighty years of age before he married.

One of the strongest passages in Harvey's Post Mortem, as showing that he accepted the popular belief with respect to Parr's age, is that in which he says, ' It seemed not improbable that the common report was true, viz., that he did public penance under a conviction for incontinence after he had passed his hundredth year;' or in other words fifty years before his death, for that is clearly all that Harvey could speak to.

But lest it be objected that I pass over without notice Taylor's remark that ' the report of this ancient man was certified to Lord Arundel, or as in his metrical narrative he expresses it

> —— by records and true certificate,

I can only observe that such records are what I have long been inquiring for in vain; and that nobody could certify of their knowledge to matters extending over a

century and a half. The most they could do would be
to state their belief. But belief is not proof; and of
anything having the semblance of proof, or the slightest
pretence to be called evidence, of Old Parr having lived
152 years, not a particle is to be found.

Parr, like Jenkins, left no children; but the lovers of
the marvellous who enriched Henry Jenkins with a son
and grandson have been equally generous with regard
to Old Parr; and not only given him a goodly number
of descendants but handed down his gift of Longevity
as a heirloom among them. For we are told that his
son lived to the age of 113, his grandson to 109, and his
great-grandson to 124; which last, Robert Parr, died at
Kinver, a small village near Bridgnorth in the county of
Salop. To these we may add Catherine Parr, his great-
granddaughter, who died in Skiddy's Almshouse, Cork,
October 1792, aged 103.[1]

A learned friend has called my attention to the manner
in which the Longevity of Parr's descendants is referred
to by Fodéré, ' Physiolog. Positif,' § 1,013, tom. iii. p. 465,
and by Nolan, ' Bampton Lectures,' p. 447; but, as Old
Parr had *no* descendants, I content myself with giving
the references for the benefit of anyone who may think
the inquiry worth pursuing—which I do not.

The fact is, Old Parr's descendants are as much matter
of fable as his 152 years of age; and I hope both will
from this time forth be eliminated from all serious inquiries respecting Human Longevity.

---

[1] ' Harl. Mis.' (ed. 1811) vol. vii. p. 69. The ' Annual Reg.' vol. iv.
p. 144, contains also particulars of another descendant : ' July 1761. Died
lately, John Newell, Esq. at Michaelstown, Ireland, aged 127, grandson
to Old Parr, who died at the age of 152.'

CHAPTER VII.

SOME forty years ago the taste of the reading public of Paris ran so strongly upon memoirs, whether of prime minister or police spies, duchesses or demireps, it mattered not, that the supply was not equal to the demand. In this emergency, when the stock of authentic memoirs was exhausted, and the public still cried with the daughters of the horseleech, 'Give, give!' the cry was answered by the compilation of a number of supposititious memoirs.

Among the most daring of these apocryphal productions may be reckoned ' Les Souvenirs de la Marquise de Crequi, 1710-1800,' a lady of whom it was pretended that she was born under Louis XIV., and lived to be presented to the First Consul in 1804. In a critique on the book in the 'Quarterly Review,' vol. li. p. 393, the late Mr. Croker exposed the utter worthlessness of the compilation, which was based on the ingenious idea on the part of the Ned Purdon by whom it was manufactured, of taking the birth of one Marquise de Crequi and the death of another, and forming from the combination his supposed Centenarian memoir writer.

What was done in Paris, advisedly and for a dishonest literary purpose, has occurred here through inadvertence and carelessness; and the improbable story of the Old Countess of Desmond proves upon examination to be the result of confounding together two or three ladies who bore that title.

Several attempts to clear up the mystery in which the identity of this lady has hitherto been involved have lately been made; but the most exhaustive and satisfactory of these is from the pen of that accomplished antiquary and genealogist, my friend Mr. John Gough Nichols, to whose paper in 'The Dublin Review,' vol. li. p. 51, *et seq.*, I am chiefly indebted for the materials of the following notes.

Nine-tenths of the forty pages of which Mr. Nichols's Essay consists are occupied with the correction of the errors and with the exposure of the false deductions of preceding writers; for what is known of the Old Countess's life may be told in very few words. A fact which has strongly recalled to my mind the remark which my venerable friend Mr. Douce made to me on the publication of his edition of Holbein's 'Dance of Death,' 'that the real history of this subject, as far as known, might be told in a moderate-sized volume; but that it would take two or three very large ones to correct the errors which had been published respecting it.'

The popular error which now prevails as to the extraordinary age of the Old Countess may, we think, be very justly attributed to the eagerness with which Horace Walpole inserted in his 'Historic Doubts,' without due inquiry, the story of her having 'been married in the reign of Edward IV., when she danced with

Richard Duke of Gloucester.' But the support which her supposed testimony gave to his theory of the personal appearance of Richard III. seems to have blinded his judgment, and he does not appear to have taken the slightest trouble to ascertain by what evidence the statement in question was supported.

Sir Walter Raleigh, in a passage to which we shall hereafter refer, mentions her marriage in the reign of Edward IV.—a gross error, as will be shown hereafter ; but with regard to the dancing, Mr. Nichols well and truly remarks: ' Though a century and a half had passed from the time when the aged Countess was finally laid in the grave, and something like two centuries and three-quarters from the days of her assumed gaiety in the English Court, yet Walpole *appears to have relied upon oral tradition alone* for this part of her history. We have searched for any printed or written record of it, earlier than his own, but without success.'

In his endeavour to identify the ' frisky old girl,' as Moore irreverently calls her, Walpole consulted the Irish Peerage, and eventually fixed upon another old Countess, Elinor, widow of the last great Earl of Desmond, slain in rebellion in 1583, who survived him till so late as 1636, having re-married the O'Connor Sligo.

Had Walpole turned to Smith's ' Natural and Civil History of the County and City of Corke,' published in 1750, he would at least have learned from the following passage who his heroine really was :

H

' 1534. Thomas, the 13th Earl of Desmond, brother to Maurice the 11th Earl, died this year, at Rath Keale, in the county of Limerick, being of a very great age, and was buried at Youghal * * * The earl's second wife was Catherine Fitzgerald, daughter of the Fitzgeralds of the house of Drumana in the county of Waterford. This Catherine was the countess that lived so long, of whom Sir Walter Raleigh makes mention in his 'History of the World,' and was reported to live to 140 years of age.'

If Walpole, whose attention had been called to the passage in question, had followed up the information here given to him, it would have saved him from falling into the many errors of which he has been guilty; but either from carelessness, indifference, or some other inexplicable motive he never took the trouble to pursue the inquiry in this direction.

In the inscription on the well-known portrait of the Old Countess at Muckross, which runs as follows, she is also described as—

CATHERINE COUNTESSE OF DESMONDE,
as She appeared at ye Court of Our Souraigne Lord King James in thys present yeare A.D. 1614,
and in the 140th yeare of her age. Thither she came from Bristol to seek Reliefe ye House of Desmonde having been ruined by Attainder. She was married in ye Reigne of King Edward IV. and in ye course of her long Pilgrimage renewed her Teeth twice. Her principal Residence is Inchiquin in Munster whither she undauntedlye proposeth (her Purpose accomplished) incontinentlie
• to return. * LAUS DEO.

When to this is added the statement that her death was occasioned by a fall from a cherry-tree or nut-tree, the reader is in possession of what is popularly believed respecting the old Countess.

Before proceeding to show what is the real history of this remarkable lady,—for that she reached a great age there is no doubt—I will quote the only two accounts of her which can be called contemporary.

The first is the passage from Raleigh's ' History of the World' (p. 66) already referred to. The book was not published till 1614, twenty-five years after the time at which he speaks of knowing her :

' I myself knew the Old Countess of Desmond of Inchiquin, in Munster, who lived in the year 1589 and many years since ; who was married in Edward IV.'s time, and held her jointure from all the Earles of Desmond since then ; and that this is true all the noblemen and gentlemen of Munster can witnesse.'

Fynes Moryson, whose ' Itinerary' was published in 1617, is the next witness in order of date, and he writes of her as follows :—

' In our time the Irish Countesse of Desmond lived to the age of about 140 yeeres, being able to goe on foote foure or five miles to the market towne, and using weekly so to do in her last yeeres ; and not many yeeres before she died she had all her teeth renewed.'

These two passages, as Mr. Nichols has pointed out, are the sources from which all subsequent notices of the Countess of Desmond are generally derived.

Let us now test these statements by the facts in the life of the Countess which recent inquiries have established.

First, as to the date of her marriage, Raleigh says
that event took place in the reign of Edward IV. So
far from this being the case, the fact is she was not
married till at least FORTY-FIVE YEARS AFTER THE
DEATH OF THAT MONARCH.

Catherine, or to use what we are told is the Irish
form of the name, Kathrin, was the daughter of Sir
John Fitzgerald of the Decies branch of the Fitzgeralds,
by Ellen, daughter of the White Knight.

The date of her birth is not known, but as she
became a mother shortly after her marriage, she was
doubtless a young woman at the time.

Her husband was Thomas the twelfth Earl of Desmond, a grandson of her great grandfather James the
seventh Earl, so that they were cousins german once
removed. She was his second wife, and as his first wife
was living in 1528 the marriage must have been subsequent to that date; for Mr. Nichols shows most
clearly (p. 69) that in 1528, the twentieth of Henry VIII.,
*forty-five years after the death of Edward IV., she was
not married*; for the following piece of evidence, communicated to Mr. Nichols by Mr. Herbert F. Hore (a
gentleman who has written much and well about the
Old Countess), proves that at that date her predecessor
Shela, was still the wife of Sir Thomas of Desmond.
It occurs in the Rental Book of the ninth Earl of
Kildare.

'Indenture from Gerald Fitz Thomas, Earl of Kildare,
unto Gyles ny Cormyk, wife of Sir Thomas Desmond,
upon Corbynere, in the county of Cork for five years,

paying 26*s*. 8*d*. yearly, and that the said Giles shall not waste the woods.'[1]

This record, as Mr. Nichols well observes, very materially affects the present inquiry. Shela, who remained the wife of Sir Thomas of Desmond in 1528, was the mother of Sir Maurice Fitz Thomas, who died in the following year. It is almost certain, therefore, that Sir Thomas did not marry his cousin, Kathrin, until after his accession to the earldom, which happened in the next year, 1529; and as in that year he granted the country of the Decies in perpetuity to Sir John Fitzgerald, Kathrin's father, it may fairly be presumed that such settlement was connected with his marriage contract. While if Kathrin was a bride in 1529, and afterwards gave birth to a daughter (who became the wife of Philip Barry Oge), it is physically impossible that she could have been born in 1464, which would be the year of her birth, supposing there was any truth in her reputed age. Earl Thomas, her husband, died in 1534; so that when Sir Walter Raleigh saw the lady in 1589 she might well be called *The Old Countess*, having then been a widow for no less than fifty-five years, during which time she had, no doubt, as Sir Walter says, 'held her jointure from all the Earls of Desmond.'

At her death, which is recorded in one of the Pedigrees compiled by Sir George Carew, Earl of Totness, where we find that 'she died in A.D. 1604,' she had therefore been a widow for 70 years ; and as she had been a mother within four or five years of the commencement

[1] Harl. MS. 3756, fol. 4.

of her widowhood, physiology and common sense point out that the Old Countess was probably about a hundred and not a hundred and forty at the time of her death.

I have taken no notice of the Old Countess's journey to London, because it is unsupported by a particle of contemporary evidence ; for I cannot regard the inscription on the Muckross portrait as evidence of that fact. To say nothing of its date, 1614, ten years after the death of the Countess, and the very year in which Raleigh published his notice of her, whose erroneous statement as to her marriage it repeats in the inscription, that very inscription appears to teem with contradictions. For while the omission of the word 'hath' (to which Mr. Nichols call attention) before 'renewed her teeth twice,' conveys the impression that the person spoken of was no longer living, in another part it is said 'her principal residence *is* at Inchiquin *whither* she proposeth incontinentlie to return' as soon as she has succeeded in the object of her visit, or to use the language of the inscription 'her purpose accomplished.'

That such a visit to the Court of England from so remarkable a person should have taken place and left no trace among the histories, pamphlets, poems or news letters of the time is scarcely possible ; and when we consider for how many years the name and fame of the Old Countess of Desmond have been the subject of literary discussion and inquiry, it is incredible that if any mention of such an event is in existence it should have escaped discovery.

But that experience has taught me better, I might
have believed in the existence of a document calculated
to throw light on her real age on the strength of a letter
to 'The Times' of May 24, 1872, in which 'A Resident
in the county of Waterford' stated it was in his power *to
confirm the statement* that the Old Countess of Desmond
reached the ripe age of 140 years, for 'a landlord in the
county of Waterford has in his possession a legal document of the time of James I., wherein it is set forth
that certain lands would fall in on the death of the
Countess of Desmond, now aged seven score years.' In
the same journal of the following day, I invited the
writer in the interest of historical truth, to furnish particulars of this extremely curious document: but, as I
anticipated, no such particulars were ever produced;
and yet with reference to this statement, which the
author could not confirm when challenged to do so, it
is gravely asserted by a subsequent writer on Longevity, that—'documentary evidence of the Countess
of Desmond's age is said to exist.'

And here I bring to a close my observations on the
three unfounded cases of supposed abnormal Longevity,
which have been so long accepted without due consideration or inquiry. I was about to speak of them as *De
Tribus Impostoribus*, but to do so would be unjust to
the Countess, who, I believe, so far from being a party
to the absurd claims put forward in her behalf, never so
much as heard of them.

In conclusion I venture to hope, that after the proofs

here given of the utter groundlessness of the claims of
these Longeval Celebrities to the fabulous ages attributed to them, future inquirers into the question of the
Duration of Human Life will no longer be dazzled or
misled by such veritable *Ignes-Fatui* as Henry Jenkins,
Thomas Parr, and the Old Countess of Desmond.

CHAPTER VIII.

I HAVE in the preceding chapters endeavoured, and I trust successfully, to show that the principles of evidence which, I contend, ought to be applied to all cases of alleged abnormal Longevity, prove that what have long been held, as the lawyers say, the three leading cases— namely, those of Jenkins, Parr, and the Countess of Desmond—must no longer be so considered ; each and every one of them being unsupported by one single atom of proof.

I now proceed to show how successfully these principles have been applied to exposing the utter groundlessness of the claims of a number of Pseudo-centenarians. I will afterwards show how, on the other hand, these very same principles support the cause of Truth, by establishing beyond all doubt the really genuine cases of exceptional Longevity.

If in furnishing the particulars of cases in which the claims of individuals to be considered Centenarians have been carefully investigated, and ultimately disproved, I commence with that of Mary Billinge, already referred to (*ante* pp. 34-7) the reader will probably exclaim with Othello, ' Why this iteration ?'

I answer it is done advisedly. In my desire to impress upon all who propose to investigate a case of Longevity the necessity of first quietly ascertaining from the friends of the old person, or the old person himself, those little incidents in life which serve to identify the individual, I have urged this necessity on two separate occasions.

So I here repeat the story of Miss Billinge by way of impressing upon my readers how the most respectable and intelligent authorities, the most earnest inquirers after truth, may be misled, from the want of experience on the one hand, and on the other, from a too ready belief in the marvellous and extraordinary; and because also there is another important lesson to be drawn from it, namely, that it is *not because a case of exceptional Longevity cannot be disproved that it is necessarily to be believed.*

Miss Mary Billinge was confidently believed to be 112; and it was just as confidently believed that this great age had been clearly proved. She was, in fact, only 91. But this could not have been ascertained had she been an only child. The fact that she had a brother and sister, whose parents were Charles and Margaret Billinge, identified her as the child of Charles and Margaret Billinge, which nothing else could have done. But for that fact, and the discovery to which it led, her 112 would have continued to exist as an element of error in all future calculations as to the average duration of Human Life.

MARY BILLINGE *not* 112 *but* 91.

The communications respecting this lady, which had originally appeared in 'The Times,' are sufficiently

indicated in the following letters, which appeared in
' Notes and Queries' of February 25, 1865 :—

' In "The Times" of January 26th, Mr. John Newton,
of 13, West Derby Street, Liverpool, communicated the
following remarkable instance of Longevity of an old
lady, whom he had attended in her last illness. The
account was written by Mr. Newton at the time of her
death, and was published in "The Times," "The Gentleman's Magazine," and other periodicals :—

' "December 20th, 1863, at her residence, Edge Lane,
Liverpool, aged 112 years and six months, Miss Mary
Billinge. She was born at Eccleston, near Prescot, on
the 24th May, 1751. She retained her faculties in a
very remarkable degree to the last, and was never
known to have been confined to her bed for a single
day until the week preceding her decease."

' On the 27th a correspondent who avowed that he
shared Sir George Lewis's doubts as to the majority
of statements of Longevity, and his wish to ascertain
the precise facts in all alleged cases, invited Mr.
Newton to furnish some particulars of the evidence
which satisfied him that the lady, Miss Mary Billinge,
who died on December 20, 1863, was the same person
who was baptized on May 24, 1751.

' I, who am also a doubter in these cases, have looked
with some anxiety for Mr. Newton's reply. That
gentleman has as yet made no sign. Parliament is
now sitting ; "The Times" will have little space for
such matters, and I hope, therefore, "Notes and Queries"
will admit an old correspondent, through its columns

to call the attention both of Mr. Newton and its Liverpool subscribers to this curious instance.

'The subject of Longevity has long attracted the
attention of men of science, actuaries, and others ;
but I believe that since the present century no case
at all approaching to that of Miss Mary Billinge has
been found to bear the test of examination.'

This elicited from Mr. Newton the following reply,
which appeared in the same journal on March 11 :—

'Your correspondent has asked me to furnish some
particulars of the evidence which satisfied me that Miss
Mary Billinge, of Edge Lane, near Liverpool, who
died on December 20th, 1863, was the same person
who was baptized on May 24th, 1751.

'In answer, I may say that it was only by a mere
accident we were able to obtain even the scanty
particulars furnished. The old lady had outlived all
her early friends. She had long been looked on as a
sort of fossil-relic of a bygone age. Her old servant,
who had faithfully served her for nearly fifty years,
died two years before herself. She was the only
depositary of the secret as to the great age of her
mistress, and, though often. questioned, she never
communicated it to anyone. But to her sister, who
succeeded to her place beside Miss ·Billinge, she told
that years ago, it had been necessary, in connection
with a will, to obtain needful certificates of relation-
ship or identity, and that Miss Billinge had then sent
her to Eccleston, near Prescot, assuring her that was
the place of her birth. We had traditional and other

evidence to the same effect. She had, it is known, a brother and sister, and she was the senior of both. The brother died in 1817, aged forty-seven years. The Health Committee in this town employed an officer to make inquiries as to the matter, who, I understood, after some research, rested quite satisfied with the truth of the certificate. Miss Billinge would never speak of the past, and always resented any reference to her great age. She had long been bent almost double with years, her skin hung extremely loose, and was most curiously wrinkled. An old lady, herself upwards of eighty years, who called to see her in my presence, looked quite fresh and youthful in comparison. Should any fresh particulars as to dates come to hand, I will communicate them.

'JOHN NEWTON.

' 13, West Derby Street, Liverpool.'

And here the matter rested until I found a friend at Liverpool who kindly undertook to pursue the inquiry. He commenced by putting himself in communication with Mr. Newton, from whom he received the following information :—   .

' 13, West Derby Street, May 30, 1865.

'Dear Sir,—In relation to the age of Miss Mary Billinge, about which you make inquiry. I have referred to my answer in 'Notes and Queries' 3rd S. vii. 207) to a correspondent's query at p. 154 same volume, and I perceive that I did not distinctly state that her baptismal register was duly consulted after her death. Her servant was sent over to Eccleston, near Prescot, where Miss B. always said she was born, for a copy of the church register. The clerk examined backwards from 90 years ago, and there was plenty of evidence, more or less certain, that she must have long passed her 90th year, until he came to the one which he copied. From this copy I gave the dates which appeared in the newspapers, and I afterwards wrote to " The Times," giving a brief account of her. It was this notice which was afterwards extracted into " Notes and Queries." She had a younger brother, whose age at death is known ; also a younger sister, who was buried at St. George's, Everton, some forty years ago. The difficulty, it appears to me, is to prove that the person named in the baptismal register was the one who died in Edge Lane. She had outlived all her early friends. She left no relative behind. Mr. Llewellyn is her executor, but has done nothing to clear up the difficulty, though I pressed him to do so. He told me, as I have mentioned in my letter, that the Health Committee sent an officer over to Eccleston, who could find no apparent error in the dates. Perhaps you might ascertain what was done. The proper plan, it appears to me, to prove or disprove the correctness of these dates would be to ascertain whether entries corresponding to the *names* of her *sister* and *brother* appear also in the register at the corresponding dates, and with the *names of the same parents*. I can furnish these if wished, and remain,

' Yours very truly,

' JOHN NEWTON.'

Furnished with all the preliminary information he could obtain, my friend set to work, and the result justified all my scepticism, as will be seen by his letter, which appeared in 'Notes and Queries' (3rd S. vii. 503):—

'I am now in a condition to furnish satisfactory information on the subject of the age of the supposed centenarian, Miss Billinge ; and I will in a few words describe the process by which I arrived at it.

'On application to Mr. Newton, surgeon, I was furnished with a copy of the certificate of baptism of " Mary, daughter of William Billinge, farmer, and Lidia his wife ; born 24th May, 1751, and christened the 5th of June.' This was assumed to be the Mary Billinge recently deceased. The question thus became one of identity. After some inquiry, I found Miss Billinge had a brother and sister buried in Everton churchyard. I have extracted the inscriptions on their tombstones as follows :—

> "William Billinge, obt. 7th May, 1817, aged 46.
> Anne Billinge, died 9th Feby., 1832, aged 59."

'I have also seen a mourning ring which belonged to the late Miss Billinge, in memory of her brother, which confirms the above date of his death. It is clear, therefore, that William and Anne were the brother and sister of the late Mary Billinge.

'The next point was to ascertain the parentage of William and Anne. I went over to Prescot church, and found the parish clerk—himself a relic of antiquity, ninety years of age, and still doing duty. He made a search for me, and found the registers of both :—

"William in 1771, son of Charles and Margaret Billinge.

"Anne in 1773, daughter of the same."

'It was clear then that William and Anne, children
of Charles and Margaret Billinge, could not be brother
and sister of Mary, the daughter of William and Lidia
Billinge.

'To put the matter beyond a doubt, I persevered in
the search, and found :

"Mary, daughter of Charles and Margaret Billinge, born 6th November,
1772, christened 23rd December."

'The identity is here complete. The old lady was,
therefore, in her ninety-first year, not in her 113th when
she died.   I suspect that most of the supposed instances
of Centenarianism will turn out to be cases of mistaken
identity.'

Mr. Newton, of whose anxiety to get the truth there
never could have existed a doubt, communicated his
satisfaction at the successful manner in which the
inquiries had been pursued, in the following letter to
the gentleman by whom it had been conducted.

'13, West Derby Street, July 1, 1865.

'Dear Sir,—I have just seen your communication to
"Notes and Queries" respecting the age of Miss Billinge,
and am glad that you have worked out the question so
satisfactorily.   Doubtless you are right.   A person at
her funeral, who though no direct relation, had in his
possession some deeds relating to the family, told me
that in these her father's name did not correspond to

that which appeared in the certificate. He also said that they calculated she was not much more than 90 years old. However, as Mr. Llewellyn and other old friends, who had known her far longer and more intimately than I, did not attach any value to this statement,[1] I said no more about it. I write now, however, to point out to you that you have cut down the old lady's age too much. If born November 6, 1772, she would be 91 in 1862, and at the time of her death, December 20, 1863, she would be 91 years, 1 month, and 14 days old, not in her 91st year, as you say. Let the old dame have full credit for her Longevity, the only thing for which she was remarkable.

'Yours very truly,

'JOHN NEWTON.'

Let me add that the tombstone of this wonderful old woman in Toxteth Park Cemetery, duly records that she was 112 years and 6 months, although under the withering influence of cross examination she dwindled into a very common-place old lady of 91 years, 1 month, and 14 days!

JONATHAN REEVES, *not* 104 *but* 80.

Happening to be at Bath when I read in 'The Times' of May 14, 1869, the following letter, I took advantage of that circumstance to look into the case.

---

[1] No. They had made up their minds that the old lady was 112; and nothing could shake that belief—which they probably still continue to hold.

I

'A Survivor of the First American War.

'Sir,—There is a man living here who was in the American revolutionary war, now a pensioner from the 62nd Regiment. He was born in the British army, and was a drummer in the same regiment with his father; he is now in his 105th year, being born in 1764.

He is in very needy circumstances, having only a pension of 6*d.* a day, and I cannot prevail upon him to go into the Union. He is living with some very kind people, but they are poor, having nothing but their labour to depend upon. It would be a great charity for any one to give him a trifle; all his relations are dead; he is not able to feed himself, he is so very shaky, otherwise his health is good. His address is—Jonathan Reeves, 5 George's Buildings, Walcot, Bath.

'I have known him for many years, and am certain this is correct.

'Respectfully, &c.,

'J. GIDDINGS.'

'8, St. James's Street, near the Abbey, Bath, May 12.'

I thought it due to the writer to call upon him in the first instance, but to my surprise found he did not reside at the address given by him.

I then called upon Jonathan Reeves; and cannot perhaps tell the story better than by giving extracts from the two letters which I wrote to 'The Times' in reply to Mr. Giddings; for in spite of his confident

assertion, ' I am certain this is correct,' I felt it was *not*,
and the result justified *my* conviction :—

'I found Jonathan Reeves, a handsome old soldier, a
little deaf, very shaky, but very intelligent, scrupulously
clean, and obviously well cared for by the good woman
who has charge of him.   He has a pension of 6*d.* a
day, which, thanks to a clergyman in the neighbourhood who receives it for him, is made up to 9*s.* a week ;
and it ought to be stated, both in justice to Reeves and
his landlady, that they were no parties to the appeal
made on his behalf.

'Reeves's memory, though clear enough as to places
and events, is very *defective as to dates.*   I could not
learn from him when or where he believed himself to
have been born, or when or where he enlisted.   The
only precise date which he remembered was that of
his discharge from the army—May 18, 1818, when he
received a pension of 6*d.* per day for 18 years' service.
But he remembered what he felt to be a hardship, if
not an injustice — namely, that three years' service
before the age of 18 was disallowed.   " I was not 18
when I fought at Maida, but I was old enough to
fight, and that, I think, ought to reckon."   He says
he was in Egypt, at Aboukir, at Maida, at Waterloo,
and after Waterloo in America ; and I suspect his
memory has a little failed him, and that it must have
been in Egypt where he was under age, and not at
Maida, which was fought on July 4, 1806.

'One thing is obvious.   If, as he states, he was discharged in 1818 with 18 years' service allowed (his

previous service being disallowed because he was then under the age of 18), it is clear that he only attained the regulation age of 18 in 1800, and consequently must have been born, not in 1764, but about 1782, and, as consequently, is not 104 but somewhere about 87.

'He is a thorough old soldier, loud in his praises of the Queen; and loud in expressions of satisfaction at the recent improvements which have been made in the condition of our soldiers, whom he describes as being now as well off as tradesmen. He inquired very anxiously about the new barracks at Chelsea, spoke warmly of the kindness of some of his old officers, especially of the late Sir Andrew Barnard, and Captain de la Bere, who is very kind to him at this time; and altogether displayed an amount of intelligence quite sufficient to prove that Mr. Giddings has been premature in adding the name of Jonathan Reeves to our list of Centenarians.'

On the appearance of this letter I was favoured with a communication from General Hutt more than confirming what I had stated. It was right that the readers of 'The Times' should be put in possession of the truth; and the following is an extract from a long letter from me respecting Reeves and other Centenarians which appeared in that journal of May 21:—

'Thanks to the courtesy of General Hutt, I am enabled to do more than confirm my former statement respecting the alleged Centenarian, Jonathan Reeves, a pensioner from the 62nd Foot. It appears that, after all, he was under age when Maida was fought,

in 1806. The records at Chelsea Hospital show that
he did not enlist before November 21, 1804, being
then 15 years of age; consequently he was born, not
in 1764, as stated by Mr. Giddings; not in 1782, as I
had inferred from his own statement; but in 1789, and
therefore is only 80 years of age, and not 104, as so
confidently stated by your correspondent. I am bound
to add that not only does this contradict his statement
that he was at Aboukir, but, further, that his name does
not appear on the Waterloo rolls.'

And so Jonathan Reeves proved to be 80, and not
104.

MARY DOWNTON, *not* 106 *but* 100.

The following account of this old lady appeared in
' Notes and Queries' of August 19, 1865 (3rd S. viii. 157).

' For the first two years after my ordination, while
curate of Allington, near Bridport, I was a weekly
visitor to a bedridden woman (a parishioner) named
Mary Downton. She died November 4, 1860, at the
(generally supposed) age of 106 years, retaining all her
mental faculties except sight; which she had gradually
lost some years before I became acquainted with her.
I can recall many a pleasant conversation with this
" oldest inhabitant." Strange to relate, the earliest
incident of her life which she could recall to memory,
was being carried out, "within an inch of her life,"
from her father's burning cottage at the age of four
years.'

This was soon after the public discussion of the case

of Mary Billinge, and the gentleman by whom it was communicated, warned by that discussion of the probability of error, kindly undertook to investigate the case as thoroughly as he could.

At first he experienced considerable difficulty in getting any precise information either as to what was the maiden name of Mary Downton or the place of her birth.

At length he succeeded ; and the result had better be told in his own words :—

' I think this case of Longevity will turn out an authentic one. Through her daughter, I learn that the maiden name of this Centenarian was Mary Hardeman, that her birthplace was Thorncombe, near Chard, and also that she was a "love-child." Accordingly, the Thorncombe Register supplies the following :—

' " Baptism in 1761. Mary, daughter of Mary Hardeman, b— b—, baptized March 22nd."

' As she died in November 1860, this would make her nearly 100 at the time of her decease ; but I well remember the old lady's telling me that she " recollected walking to church to be christened at about the age of four or five years." She may therefore very fairly have been in her 105th year at the time of her death, which is only one year less than the age which she claims to be.'

On this I must be permitted to remark that children have no doubt frequently walked to church to be baptized, some even ' in pattens,' as old Mrs. Puckle fancied she recollected doing ; but we suspect that ' love-children ' are most frequently baptized when the mothers

are churched; and under these circumstances we feel
bound to limit our belief to what this certificate confirms, viz., that old Mary Downton was nearly a
hundred at the time of her decease.

JOSHUA MILLER, *not* 111, *but* 90.

The case of this old man is very interesting, and very
instructive in many ways. Of the good faith with which
it was brought forward there cannot be the smallest
doubt; and that, too, by gentlemen of position and
intelligence, and not carelessly; nor until after what
was believed by them to be a thorough and satisfactory
investigation.

While, as the reader will see in the end, it shows how
hard it is to disabuse—I will not say, popular belief—
but the belief of right-minded men, and intelligent
people, when they have once allowed it to take root in
their minds.

The story, as I must tell it, will be a rather long one,
but, I trust, not longer in the opinion of my readers
than the circumstances of it fully justify.

My attention was first called to the case by an article
in 'Notes and Queries' of July 1871, where it was
stated that 'There is now living in the workhouse at
Morpeth, in Northumberland, a man aged 110, who,
until very recently, might be seen walking about the
town, and in possession of all his faculties. He is a
native of Whickham, where his baptismal register has
been sought for and found. He was in early life seized

by the press gang, and served for some time under Nelson in his own ship. There is a photograph of this old fellow, and a comely portrait it is. His name is Joshua Miller.'

I was then enabled, by the kindness of a mutual friend, to put myself in communication with the Hon. and Rev. Francis Grey, Rector of Morpeth, who had taken great pains to inquire into the truth of the old man's alleged Longevity; and had, as well as Lady Elizabeth Grey, shown considerable kindness to the poor old fellow. From him I received, in addition to many other interesting particulars, what was considered to establish Miller's age beyond the possibility of doubt—a copy of his baptismal certificate, which showed that ' Joshua, the son of Robert Miller and Ann his wife, was baptized at Whickham, the 25th October, 1761.'

Moreover, by the kindness of my friend, Mr. Woodman of Morpeth, I was put in possession of copies of two reports upon Miller's case, made by Dr. Paton, the medical officer of Morpeth Union. One of these, 'On the Personal History of Joshua Miller, as gleaned from himself and his daughter,' deserves to be printed at length. If all gentlemen in a similar position to Dr. Paton would exhibit the same intelligence and take the same pains when cases of supposed Centenarianism come under their notice, they would render good service to Human Physiology.

The following is the report referred to :—

' Joshua says he was born at Teams, on the Tyne ; his father's name was Robert Miller, and his mother's

Thomasina Coates. His father was a keelman, and lived to the age of 114 years. He himself worked as a keelman until he was pressed into the Royal Navy at Shields, but in what year or at what age he cannot remember. He was drafted into the war frigate *Pomona*, commanded at that time by Captain Lobb, who, he says, was the oldest captain in the navy. The ship cruised principally off the coasts of France, Portugal, and Spain, during the *five* years he was on board of her. He was not discharged, but left the service of his own accord. He talks of an interview he had with Lord Nelson at Portsmouth, and takes credit for being the means of doing away with flogging in the navy.

'He next turns up as working a keel upon the river Blyth. He was married twice; first, at Horton Church, but in what year or at what age he cannot tell. Eight children were the issue of this marriage. His wife died, but he cannot tell when, and he lived a widower for several years. He married his second wife at Bedlington, but in what year or at what age he cannot remember. He had issue by this marriage, one daughter, now forty years of age.

'This daughter, now Mrs. Cockburn, lives at Stockburn, and is the mother of nine children. She is a remarkably strong, powerful woman. She says her earliest recollection of her father is that of an old man (he would then be 70) with a bald head, and long white hair at the back part of it, with which, as a little girl, she used to play and try to curl. He was uniformly healthy, and was a very large eater, particularly of

animal food. He could neither read nor write; was
not by any means an intelligent man, but full of old-
world stories, which were generally laughed at. He was
happy and contented, and cared for nothing but his food
and his " 'baccy ;" in other respects he was a temperate
man.

'Her mother died thirteen years ago, aged 63. She
says an aunt, a sister of her father's, died a few years ago
at Gateshead, aged 108 years. This statement the old
man confirms.

'The maiden name of her father's first wife was
Isabella Pringle, and that of the second (her mother)
Mary Turner.

'R. PATON.'

'Morpeth, August 31, 1871.'

Dr. Paton's judicious inquiry had been the means of
eliciting three very important facts.

In the first place the old man stated distinctly that
he was born, not at 'Whickham,' but at 'Teams on the
Tyne.'

Secondly, what was no less important, that his
mother's name was not 'Ann,' but Thomasina—
'Thomasina Coates.' I pointed out the discrepancy, as
tending to show that in all probability the Joshua Miller
of Whickham, the son of Ann Miller, was a different
person from the supposed Centenarian. But I was met
with the objection that Ann was doubtless a shortening
of Thomasina, a name which no keelman would think of
using. The reply to this objection was obvious—

namely, that in the case before us, the abridgement of
the name, if any, had been made by the officiating
clergyman ; it was the keelman himself who said that
his mother's name was not Ann but Thomasina.

The third point ascertained by Dr. Paton was one
calculated to throw a good deal of light on Miller's
history. It was to the effect that he had worked as a
keelman until he was pressed into the Royal Navy
at Shields, *but in what year and at what age he could
not remember*, and that he was drafted into the war
frigate 'Pomona,' commanded at that time by Captain
Lobb, who, he said, was the oldest captain in the navy.
The ship cruised principally off the coasts of France,
Portugal, and Spain during the five years he was on
board her ; that he was not discharged, but *left the
service of his own accord.*

Here was a plain, straightforward statement of facts
which it was possible to test.

Having ascertained from the Navy Lists that Captain
Lobb commanded the 'Pomona' in 1805 and 1806, I
ventured to ask for any information which the books of
the 'Pomona' of that time might contain respecting
Joshua Miller.

The name of *Joshua* Miller was not to be found, but
there was a *Joseph* Miller, and there can be no doubt
that this was a clerical error for Joshua, and that he was
identical with the old man in Morpeth Workhouse—
inasmuch as the man was born at Newcastle, joined the
'Pomona' in 1805, and '*retired* from it at Lisbon, November 30, 1805.' This latter fact confirming, as it does,

Miller's statement 'that he left the service of his own accord'—a practice, I suspect, very rarely permitted in those days at least—completely identified the subject of the inquiries. But the ship's books tell us in addition that when Miller joined the 'Pomona' in 1805 he was 22—that is, was born in 1783, and not in 1761— and was consequently, in 1871, not 90 years of age, much less 110 or 111.

Here, after some fruitless attempts to discover the right baptismal certificate of Joshua Miller, I was content to leave the case, satisfied in my mind that the official record was to be depended upon—a satisfaction, I think, not fully shared by all those who had believed the old fellow's exceptional age.

In January last there appeared in 'The North of England Advertiser' a paper written by one who styled himself 'An Old Stager,' on 'Centenarianism in the North of England,' and in this paper the story of Joshua Miller is retold, as will be seen, with some rather telling additions :—

'In some respects it is pleasant to know that, after descanting upon the wonderful length of existence of those named, we have still in our midst one who is hale and hearty, and who has reached the patriarchal age of one hundred and ten years—namely, Joshua Miller. This old man fought under the gallant Nelson, and, like all old tars, his greatest glory is in repeating the deeds of the bold and the brave—how when the 'Pomona' frigate was once lying off Boulogne, a live shell was thrown on deck, which he fearlessly grasped and pitched

overboard—how he narrowly escaped kingdom-come
seventy or eighty years ago—how he laughed when his
enemies went by the board—how that his father lived
to the age of 119, his sister to 120, and that he hopes to
do the same, with an endless variety of characteristic
yarns which would lose their pungency by repetition.
About twelve months ago he "astonished" the officers
and captain of the gunboat "Castor" by the agility with
which he paced the deck, scanned the rigging, and dived
into the cabins.   It is affirmed that Joshua was born at
Gateshead, in 1761, and the record of his baptism is in
the parish register at Whickham.   If this statement
is true, we have another refutation of the assertion that
no person "can be proved to have lived 100 years."   A
couple of years since, Mr. Geo. Grey, assistant overseer, Ridley Villas, kindly endeavoured to raise a subscription to keep the hearty old soul off the parish, or
out of the poor-law union, but I believe the good in-
tentions of Mr. Grey were not successful.   At any rate,
by a paragraph in the papers, I observe that Joshua
Miller is an inmate of the Morpeth Workhouse, where
he partook of a hearty meal on Christmas-day.   This
appears to be a lamentable conclusion to the old tar's
career, and almost to imply that the old fellow has
"braved the battle and the breeze" to small purpose.
So far as is convenient in such establishments, it is
satisfactory to know that the veteran is well treated
and comfortable—that his appetite and digestive organs
are in full trim—and that not one of those houseless
wanderers enjoyed his dinner with greater zest than

Joshua Miller. If life is worth having, and even small pleasures only now and then shed their radiance, may it be many a long year before they " rattle his bones over the stones." '

The reader who has just seen that Miller only entered the ' Pomona ' in 1805, and retired from it in November of the same year, will admit that in the way of picking up adventures he made very good use of his time.

I refrained from pointing out the many errors in this statement, being unwilling to share the imputation which Dryden cast upon Alexander,

> ' And thrice he routed all his foes,
> And thrice he slew the slain,'

and determined to say what I had to say in the present book.

But in April, 1872, poor Miller, died and his death was recorded in ' The Newcastle Daily Journal' of the 25th of that month as follows :—

' There died yesterday, in the Union Workhouse, Morpeth, an inmate, Joshua Miller, said to be 111 years of age. He was a native of the Teams, and was baptized at Whickham Church in October, 1761, the register of the performance being extant. In early life he was a keelman on the Tyne. He was pressed during the great Continental War, and served on board the ' Pomona ' war ship while Nelson was achieving his famous victories. Latterly, he was employed at Bedlington Iron Works, in conveying the manufactured goods down the river to the port of Blyth. He became an inmate of the workhouse about two years ago, and

was able to move about till within a week ago. He
was among the inmates in the dining room when they
were entertained to tea last Easter by Lady Elizabeth
Grey. His departure from this life was announced to
the town yesterday by a muffled peal. He is to be
interred on Saturday.'

   But Truth is sure to prevail sooner or later; and it is
only just to the gentleman who contributed the curious
Centenarian records to the 'North of England Advertiser' to say, that in that journal of May 4 he told the
good people of the north how old Joshua Miller really
was :—

   'In letter No I., which appeared on the 12th of
January last, a notice was given of Joshua Miller as a
living wonder, when his age was stated to be 110 years ;
that his baptismal register was at Whickham ; that
he had fought under Nelson ; that he had picked up a
live shell when on board the "Pomona" frigate and
threw it overboard ; that his father lived to 119 and his
sister to 120, and he hoped to do ditto; that he
astonished the officers and men of the gunboat "Castor"
at Shields by his agility ; that he was at that time in
Morpeth Workhouse ; that a subscription was raised
in his behalf ; concluding with a wish that it might be
long before the hearse "rattled his bones over the
stones."   Well, Joshua Miller died on Wednesday,
the 24th of April, 1872, and was buried from Morpeth
Workhouse on Saturday, April 27th.   In his sphere as
a pauper, the old man received deference from all, was
placed in the van of his compeers, and honoured with a

muffled peal on his demise. There is no denying but
these extra civilities were in great part due to Joshua
being considered a sort of curiosity at 110 years of age ;
but, alas ! twenty additional years at the fag end of a
man's life (particularly when he is within hail of a
hundred) make an immense difference. Though the
kindness of Mr. George Grey, assistant overseer for the
parish of All Saints, Newcastle, I am able to correct previous inaccuracies. It happened that Mr. Grey was at
Whickham on business—this was after an appeal was
made to the public, and money had been received—and
resolved to settle the disputed point as to Miller's age.
The old man, or his family, had previously obtained a
copy of his register, which stated that "Joshua Miller was
baptized as the son of Robert and *Ann* Miller, in the
year 1761." Mr. Grey, however, inquired of the supposed
Centenarian if he knew his mother's name, when he un-
hesitatingly replied, "Thomasine." With this hint
the enigma was at once unriddled, by Mr. Grey and an
official at the church reading, that "Joshua Miller, the
son of Robert and *Thomasine* Miller, was baptized at
Whickham, in the year 1783." Thus, instead of 110,
the Joshua Miller who died last week could not have
exceeded his ninetieth year. Mr. Grey acquits the
veteran of any wish to deceive his patrons, and Miller
really believed he was as old as his first register represented ; even the rector of Whickham was misled, and
helped the subscription list with a donation. Further
confirmation was received by Mr. Grey from a man at
the Teams, close upon 85 years of age, who was a

schoolfellow of Miller's and knew that the latter was only about four years his senior when a boy, "unless he had gathered-up an extra twenty years or so when an old salt." "But really," Mr. Grey kindly concluded, "when all is said and done, poor Joshua was a wonderful man for his years, remarkably fresh, and deserved a better fate than the workhouse at last."'

And so, as I had contended, the old man was right as to his mother's name; and, as I had contended, the books of the 'Pomona' gave us his real age; and when he died poor old Joshua Miller was not 111, but just about 90.

MAUDIT BADEN, *not* 106¼:

It was while at Bath, at the time I investigated the case of Jonathan Reeves, that my attention was called to the following paragraph, which appeared in 'The Times' of May 13, 1869:—

'There died on Tuesday morning, May 11, 1869, at Oare, in the parish of Wilcot, Wilts, Matthew Baden, at the age of 106¼ years, having been born at Pewsey in the month of February, 1763. He has left a numerous offspring of children, grandchildren, great-grandchildren, and great-great-grandchildren. His eldest daughter, now living, is verging on 80 years, and has children and grandchildren many. Matthew Baden did not marry till after he was 30 years of age. Till latterly he cultivated a few acres of garden land, and he died respected in his own little freehold cottage at Oare, after a confinement to his bed of four days only.'

This paragraph went the round of the papers, and I made up my mind on my way back to London to visit Oare, and examine on the spot the truth of the 106¼ years attributed to this Wiltshire Patriarch.

Circumstances prevented me from carrying this arrangement into effect; and when I began to investigate the case I was met at the outset by the too frequent impediment, namely that the baptismal register was defective for the period when Baden is said to have been baptized; and the tradition as to how the injury was effected is certainly more amusing than satisfactory.

It is said that a former incumbent having gone to the church to marry a couple, accompanied by a favourite greyhound, the dog was shut up in the vestry while the ceremony was being performed, and amused himself by tearing out several of the leaves of the register, and among them the one on which Baden's baptism is supposed to have been entered.

Some time after this a clergyman of the neighbourhood, to whom I was introduced by a mutual friend, kindly undertook to look into this case for me, and the result was just what I anticipated.

In the first place there is no evidence as to the date of either the birth or baptism of the old man, whose name by the bye is not Matthew but Maudit or Mardit Baden. His eldest daughter Martha, who is described in the paragraph quoted as having been 'verging on 80' in May 1869, turns out to have been born April 9, 1798, so that she was then just 71.

I have not been able to ascertain what authority

there is for the statement that Baden was not married
'till after he was thirty years of age.' His son, who is
living, was questioned upon this point, but his only
answer was that he 'never 'eared vaather zay nout
about it.' But if Baden was more than 106 when he
died he must have been *much* more than 30 when
he married, as that took place in January 1798. But
I have no doubt the real truth is that which is asserted
on very high authority in the neighbourhood, that 'he
jumped on ten years when about 70,' or as it is put by
one who knew him, 'he slipped on ten years a long
while ago,' and that after all Mardit Baden, instead of
being half-a-dozen years more than a century, was
really some few years less. Nevertheless those who
reported his death to the registrar reported him as 106,
and as 106 he figures in the Registrar-General's annual
report for 1869 ; for, as has been already observed, the
Registrar-General has no alternative in these cases, but
to tell the tale as it is told to him.

THOMAS GEERAN, *not* 106 ?

I am sorry to enter at such length as I have felt right
into the exposure of this most impudent case. But the
unwarrantable persistency with which it was from time
to time brought forward, demands a thorough exposure ;
while in addition, it affords a striking illustration of the
difficulties which ignorance and prejudice throw in the
way of those who are desirous of ascertaining the truth.

On May 28, 1870, a valued contributor to 'Notes

and Queries' (4th S. v. 522), in giving some information respecting Jane Scrimshaw, an alleged Centenarian, concluded her reply with the following notice respecting Thomas *Guerin*, who, when the announced life was published, was called *Geeran* :—

'As many correspondents of "Notes and Queries" take an interest in the question of Centenarianism, the enclosed announcement will no doubt interest them. May we accept Mr. Guerin as a living refutation of those scholars who, like Mr. Thoms and the late Sir G. C. Lewis, doubt the possibility of a human being living a century ?—

' " To be published by subscription, 1*s.* to subscribers ; non-subscribers, 1*s.* 6*d.* A Sketch of the Life of Thomas Guerin, the Brighton Centenarian, being an Answer to the late Sir Cornewall Lewis, on his Theory of Longevity, by R. H. Williams, M.A., Ph.D. (Lecturer on Chemistry and Natural Philosophy, Author of " Charsley Hall," &c. &c.) With a photograph of Thomas Guerin, by M. Lombardi of Brighton.

' " The list of subscribers will be published in the work, and the proceeds go for the benefit of the old man and his wife.

' " Thomas Guerin, who is now in his 104th year, was present at the capture of Seringapatam, in 1799 ; at Corunna, in 1809, he received two gun-shot wounds ; at Vittoria, a severe sabre cut in the head. He escaped through Waterloo, and entered Paris with the victorious army ; was discharged, invalided, from the 71st Highlanders, in 1819, with 114 days' pay, but without any pension, and is now, as may be supposed, in very straitened circumstances.

'" Sir William Verner, Bart., who had been in the same campaigns with Guerin, has sent him 5*l.* through Dr. Tuthill Massy, of 17, Denmark Terrace, Brighton, who will be pleased to receive subscriptions towards the above object.

'" Names of subscribers will be received by the London publisher, Robert Hardwicke, 192, Piccadilly; Mr. T. M. Feist, the Circulating Library, 80, King's Road; and at 'The Advertiser' office, 19, Middle Street, Brighton."'

When I procured a copy of this life, which, as a specimen of rigmarole, I did not think could be exceeded—an opinion which I had to modify when I read the second edition of it—I applied to it the late Mr. John Wilson Croker's favourite test, compared the dates, and was soon convinced that Geeran's statement was nothing better than a tissue of falsehoods; an opinion quite confirmed when I found that it had been carefully investigated by the authorities of Chelsea Hospital, and that not one of his statements could be authenticated by the records of the Hospital or of the War Office.

I am sorry now that I allowed the contempt I felt for the case, coupled with the fact that I was much occupied at the time, to interfere with my publicly exposing it.

So the matter rested until November 20, 1871, when, on the death of the old man, a correspondent, H. P., sent the following account of him to 'The Times':—

'I send you the following particulars of the life of a veteran, Thomas Geeran, who died a few days ago in the infirmary of the Brighton Union, at the advanced age of 105 years.

'These particulars I gathered from the lips of the old man himself; and from the inquiries which I have made, I have every reason to believe that they are reliable. Should you deem them worthy of record, I should feel obliged by your inserting them in "The Times."

'Thomas Geeran was born on May 14, 1766, at Scarriff, in the county of Clare, Ireland. Bred a sawyer, he continued to work at his trade till the year 1796, when he enlisted in the 71st Regiment.

'He shortly afterwards went out to India, and on the breaking out of hostilities with Tippoo Sahib, the Sultan of Mysore, he was engaged in the siege of Seringapatam, which was carried by storm on May 4, 1799.

'On his return to England he accompanied his regiment on the expedition to Walcheren, and thence, after a short interval, he proceeded with large reinforcements to join the army of Lord Wellington in the Peninsula, where he continued to serve until the conclusion of the war, in April 1814.

'He was engaged in almost every battle fought by our army after our arrival in the Peninsula, and was severely wounded on three occasions.

'On the return of Napoleon from Elba he was sent with the army under the Duke of Wellington to Belgium, and concluded his active military career at Waterloo; and in 1819 he was discharged from the service.

'For many years he gained his livelihood by working at his old trade as a sawyer, and when he became infirm he contrived to eke out an existence by the contributions of the charitable, among whom, I believe, may be reckoned her Majesty the Queen and several other members of the royal family.   He retained the perfect use of all his senses to the last, and his memory, which was wonderfully retentive, remained unimpaired to the end of his life.'

I felt bound to answer this statement, and the following is my reply which appeared in ' The Times ' of the 22nd of the same month.

'November 22, 1871.

' Sir,—I hope your correspondent " H. P.," who has sent you an account of " Thomas Geeran, who died lately in the infirmary of the Brighton Union at the advanced age of 105 years," will forgive me for pointing out a very important omission in his letter.   It does not contain a particle of evidence in support of any one of Geeran's statements.   True, " H. P." says, — " From inquiries I have made I have reason to believe these particulars are reliable ; " but he does not tell us where these inquires were made.   Were they made at Scarriff, where Geeran was born, or at the War Office, or Chelsea Hospital ?   I doubt if they were made at Scarriff ; I am sure they were not made at Chelsea.

' Geeran's story was that he enlisted in the 71st in 1796, that he served in India, in Egypt, in the Peninsula, and at Waterloo, and was discharged in 1819 (after a service of 23 years, including India) without a pension !

' Now, I do not believe there is one atom of foundation for this story.

'In 1868 Geeran's case was brought before the Commissioner at Chelsea, when every endeavour was made to verify it; but it vain. His name does not appear on the Prize Rolls for the Peninsula or Waterloo; neither does it appear on the Medal Rolls, though Geeran stated that he had received the Waterloo Medal.

'Any one who knows the accuracy with which these official records are kept will be satisfied that his story about the Peninsula and Waterloo is groundless: and with this official evidence of their falsity, who can believe the other unsupported statements of this "old soldier"?

'A life of Geeran was published at Brighton in 1870, in which there exist several inconsistencies, and what I cannot but think a great want of candour. The result of the investigation into Geeran's case at Chelsea Hospital two years before is all but suppressed; for surely the belief expressed, that a "pension could have been obtained for him had the clerks succeeded in finding his name on the books," can scarcely be considered as a fair statement of the fact that his case had been thoroughly investigated at Chelsea with the result I have just shown. I have before me four different photographs of Geeran, taken on what he said was his 104th birthday, and I feel convinced that any physiologist would at a glance pronounce them to be portraits of a man nearer 80 than 104 years of age. The indications of extreme age which are so marked in the portrait of a genuine Centenarian are entirely wanting in these admirably executed photographs of Thomas Geeran.'

I was not surprised to find that the communication evoked replies from the believers in the abnormal Longevity of old Geeran. Accordingly ' H. P.' replied to me as follows, through ' The Times :'—

' Sir,—With reference to the particulars I sent to you respecting the late Thomas Geeran, and the letter of your correspondent on the subject in ' The Times ' of to-day, I trust that you will allow me to point out to him that I did make inquiries at Scarriff, through Father O'Malley, the parish priest of that place, when I was staying in the neighbourhood in 1867, and that the result of that inquiry was that a man then in Scarriff, named Geeran, aged 72, informed Father O'Malley that he had an uncle, named Geeran, who had enlisted in the army when he himself was only two years old, and that the family had never heard of their relation since. This evidence would seem to tally with Geeran's own statement as to the date of his enlistment. That the fact of his not having a pension was owing, as he admitted himself, to his having been discharged from the service for misconduct ; that the same cause will account for his not having the Waterloo medal, which, under the circumstances, became forfeited, and also deprived him of any claim to the Peninsula medal when subsequently issued ; that from the many conversations I had with the old man, from the accuracy of his statements with regard to dates, places, persons, circumstances, &c., and from the many ways in which I tested the veracity of his tale, there was scarcely room—indeed, to a soldier, no room—to doubt that he had been a soldier for many years.

'It does, however, occur to me that, like many more, he may have enlisted under an assumed name, and, if so, all trace under the name of Geeran in the official records of the War Office would be lost.

'While concurring in your correspondent's remarks with regard to the so-called "Life" of Geeran, published at Brighton last year, I must demur to his idea of the age of Geeran as derived from photographs. I never in my life saw a man more bent with age.'

And in 'The Times' of the 25th of the same month Dr. Tuthill Massy, of Brighton, favoured me with the following few remarks on my incredulity :—

'Sir,—Having made a *post-mortem* examination of Thomas Geeran, I cannot let Mr. Thoms' letter pass without a few remarks on his incredulity, which, with your permission, I shall briefly state in reply to Mr. Thoms' evidence, a part of which runs thus :—

'"I have before me four different photographs of Geeran, taken on what he said was his 104th birthday, and I feel convinced that any physiologist would at a glance pronounce them to be portraits of a man nearer 80 than 104 years of age."

'I have compared these photographs, and it is surprising how unlike each other they are. One looks twenty, I may say thirty years older than the other three, although taken on the same day; but let Mr. Thoms speak on. He says :—

' " The indications of extreme age, which are so marked
in the portrait of a genuine Centenarian, are entirely
wanting in these admirably executed photographs of
Thomas Geeran."

'I have the photographs of "two genuine Centenarians"—Mathew Greathead, of Richmond, in Yorkshire ; the other, Richard Purser, of Cheltenham, aged
112. Their likenesses are both younger-looking than
Geeran's, although their heads are larger ; Geeran had
a small, compact head, very well formed, and his features
were small, thus leading a superficial observer to picture youth where real age existed.

'Again, it appears childish asking for the registered
birth of a publican's son in an obscure village in a
remote Irish county, when it is known to Mr. Thoms,
as stated in the Introduction to the "Life of Geeran,"
that :—

' " In those distant days registration was not much
thought of by doctors or divines ; so careless were even
the 'noble family' of His Grace the Duke of Welling-
ington, that even to this hour it is a question of inquiry
whether he was born in Dangan Castle, county Meath,
or in the city of Dublin. Therefore our readers will
have to accept traditional testimony. founded on local
and historical events as evidence sufficiently convincing
for minds capable of receiving circumstantial evidence."

' Several gentlemen in Brighton believe in the truthfulness of Geeran's statement—one, I may mention,
whose soundness of mind and knowledge of men and
books are a sufficient guarantee for his judgment. Mr.

George Long knew Geeran for ten years, and at the end of that period said :—

' " I entered his age when first he called on me, and since, on each birthday, he has repeated the same, adding one year correctly. Now, at the end of ten years he appears younger than when I first knew him, looking then a very old man."

' Mr. Long also testified to Geeran's love of reading translations from classical authors, and when questioned remembering what he had read.

' Mr. Thoms refers with great confidence to the inquiry made at Chelsea. This I can answer by stating there is an " old soldier" in the Brighton Workhouse who applied in vain for forty years, and at the end of that time had his name discovered and got his pension.

' Mr. Thoms is really not justified in saying " I do not believe there is an atom of foundation for (Geeran's) story." From a knowledge of Mr. Thoms's opinion I am convinced that had Geeran's birth been registered, Mr. Thoms would have said " Oh, it is not correct ; this registration is Geeran's grandfather, Old Tom, after whom Young Tom was named."

' Geeran accounts for his having foolishly enlisted at the ripe age of 30. He held the appointment of clerk in the office of a wealthy firm in Waterford, and was raised to an advanced post as agent to the branch house in America. Before starting he joined some acquaintances to have a jolly farewell, got drunk, and enlisted, which to the last he mourned.

' The examination of Geeran's remains has revealed as

much of the mysterious as his eventful life. This will appear in the " Medical Times and Gazette." Suffice it to say that not one of his medical or lay friends suspected the amount of disease within the old soldier. How he could have battled through life, and ascended the Brighton hills for years, out in every weather, with a heart closely bound up in the pericardium, a lung closely adhering to the ribs, and cancer of the pyloric orifice of his stomach, is far more difficult of comprehension than the useless folly in those who question his age.

> ' I am, sir, yours faithfully,'
>
>   ' R. TUTHILL MASSY, M.D.
>
> ' 17, Denmark Terrace, Brighton, Nov. 23.'

My rejoinder to these communications appeared in ' The Times ' a few days after, and brought this correspondence to a close :—

' Your correspondents " H.P." and Dr. Massy, who has made a *post-mortem* examination of the body of Thomas Geeran, still believe the story that he was 105 years and 6 months old.

' I, on the other hand, am convinced that the story of Geeran's life, as told by himself, is entirely without foundation—as I think I can convince your readers.

' Before doing so, let me remind " H.P." and Dr. Massy that, in defiance of the great rule of law and common sense, " *Illi incumbit probatio qui dicit, non qui negat*," the proof lies with him who makes the statement, not with him who denies it — neither of them has produced one *iota* of evidence in support of Geeran's extraordinary story; whereas, just in proportion as the facts alleged are exceptional and contrary to general experience, they ought to be accompanied by proofs direct, clear, and beyond dispute.

'Now, Geeran states he served twenty-three years in the 71st Foot, was with his regiment in India, Egypt, the Peninsula, and at Waterloo, and was discharged from the army in 1819 without a pension.

'Now, though his case has been thoroughly investigated by the authorities on no fewer than three separate occasions, his name is not to be found on the regimental rolls or on the prize or medal rolls for the Peninsula or Waterloo. "H.P." tells us Geeran admitted that he had been discharged for misconduct, on which account, says "H.P.," his claims to the Peninsula and Waterloo medals and prize money would be forfeited. "H.P." has been misinformed on the point. I have the best authority for saying that in framing the prize and medal rolls, all that is required is that the men whose names are there entered should have been present with the army at such actions or captures, and that, so far from character being considered, there are the names of men recorded on the prize rolls who were "convicts" when the rolls were framed. But even if this had not been so, Geeran's misconduct could not have caused his name to disappear from the records of the 71st Foot, and his name is not there. In fact, that he should ever have served as he stated, and no trace of such service be discoverable is simply impossible.

'But just let me call attention to the contradictions in the story he tells of his enlistment. He was born, he says, at Scarriff in 1766, and remained at school till he was 20—that is, till 1786; lived at home two years, till the death of his father (1788), then removed to Waterford, where he found employment, and at the end of three years, when under the influence of drink, he enlisted into the army.

'Then follows a very confused account of his doings, until he landed at Madras after a voyage of a year and two days, in 1797. Yet in the latter part of his narrative he is twice made to state that he enlisted "at Waterford in March, 1796."

'"H. P." objects to my drawing any inferences as to Geeran's age from his photographs. If I am wrong in my estimate of the value of photographs as evidence in such cases, I may plead that I share the error in common with one of the most eminent men of science in Europe.

'One word as to Dr. Massy's "Autopsy of a Centenarian." If, disregarding for a moment the caution, "*Ne sutor*," &c., I refer to the mysteries of medical science, I would merely remark that, having read the "Autopsy," and found from it that Geeran died of disease, and that, "death did not result from old age," "that length of years did not lead to his death," "that he had a remarkably quick blue eye, without a trace of the *arcus senilis*, and no ossific deposit in his cartilages," I think I am justified in drawing the conclusion that common sense, photography, and dissection unite in proving that Thomas Geeran was no more 105 years old than I am.

' I remain, sir, yours very faithfully,

'WILLIAM J. THOMS.

'40, St. George's Square, Nov. 24.'

' P.S.—The foregoing was written before I read Dr. Massy's letter in "The Times" of this morning, which, as it avoids the great point at issue between us—viz., what evidence there is of Geeran's age, scarcely calls for a reply. I have the photographs referred to. Mat. Greathead, said, but not proved, to be 100, looks, in my opinion, much older than Geeran. Richard Purser, —to whose 112 years there is no other evidence than his own recollections, and what a lady now living has heard —looks much nearer 80, as I believe he really was. If " H.P." or Dr. Massy would compare any of these photographs with the vignette portrait of Mr. Luning, taken by Mr. Dawes,[1] of Blackheath, after that gentleman had completed his 100th year, I think they will admit the value of photographs in inquiries of this nature."

The results of the *post-mortem* examination of Geeran were published in ' The Medical Times ' of November 25, 1871, under a title which, looking at the evidence which had been produced, was, I think, quite unjustifiable—' Autopsy of a Centenarian.' On this report I, as a non-professional man, am not qualified to sit in judg-

[1] This should have been ' Mr. Buchanan Smith.'

ment. I pass it over, therefore, with the remark that Dr. Massy admits that Geeran's death was *not the result of old age,* and that *length of years did not lead* to his death so much as force of will, and that ' had I not been told of this.force of will, I should have supposed the old fellow had died from some of the several probable causes of death described in the autopsy.'

About the same time the second edition of the ' Life ' was published, in which Dr. Massy made himself very merry at my expense, though I cannot but think that in a work written by a professional gentleman, on a question of considerable scientific importance, some of the mirth might with advantage have given place to a little closer argument and a little more candour.

But the absurd story with which the old soldier and his credulous supporters had so long beguiled the good people of Brighton at length received its *coup-de-grâce.* The result of a thorough official inquiry into the case was placed in my hands ; and I printed it in ' Notes and Queries' of March 2, 1872, feeling as I there said that ' so exhaustive and complete a demolition of the series of falsehoods by which Geeran had imposed upon the benevolent deserved to be published without alteration or abridgement.'

' THOMAS GEERAN AN IMPOSTOR.

' Remarks on the Statements contained in a book called " Longevity : The Life of Thomas Geeran, late of the 71st Highlanders."

' Determined, if possible, to fathom the mystery of

L

this old man's reputed services in the 71st, I went to the Public Record Office, and obtained access to the original muster rolls, pay sheets, and description roll of this regiment, for a period extending from 1780 to 1830, which period more than covered the time of his alleged service.

'From this search I extracted the following information :—

'In 1796, the year of his alleged enlistment, there was no such man on the pay-sheets of the 71st, nor was there any name at all like it.

'In 1799, the year alleged in which he was present with the 71st in India, there was no such man or name on the pay-lists of the regiment.

'In 1801, the year when he alleged he was in Egypt, there was no such name on the rolls.

'In 1809, the year Corunna was fought, at which battle he alleged he was present, there was no such name on the rolls.

'In 1815, the year Waterloo was fought, at which battle he alleged he was present, there was no such name on the rolls.

'In 1819, the year in which he alleged he was discharged, there was no such name on the rolls.

'It may fairly be asked then, is it possible that he could have served as he alleged, and yet not have his name on these rolls? The pay-lists are the originals forwarded quarterly by the paymaster, and containing the name of every member of the regiment drawing pay, and therefore fully to be relied upon.

'Where, then, could this old man have picked up all his wonderful anecdotes and asserted reminiscences of the exploits of the 71st? The following information will, I think, go a long way to prove who this man really was, and why he should have picked out such a regiment as the one he did.

'It appears from the pay-sheets of the 71st Foot in 1813, that there was a man of the name of Michael *Gearyn* or *Gayran*, then serving.

'From the description roll it appears that he enlisted *March* 3, 1813, and *deserted* on *April* 10, 1813.

'He was born at Turlee (*sic*) in the county of Kerry, Ireland, and was by trade a tailor. The following is a comparative description of Thomas Geeran and Michael Gearyn, by which it will be seen that in appearance, &c. there must have been so great a resemblance between these two men as almost to establish their identity :—

'Thomas Geeran [1], born at *Tulla, Killaloe, Clare*; height on enlisting, 5 feet 10¾; hair, white in 1870; eyes blue; complexion fresh.

'Michael Gearyn, born at Turlee (?) co. Kerry; height on enlisting 5 feet 9¾; hair brown; eyes blue; complexion fresh.

'Thomas Geeran, when asked the name of the officers of the regiment, could only recollect two, Col. Denis Packe and Lieut. Anderson the adjutant.

---

[1] Thomas Geeran stated his father's name was *Michael*. This account of his personal appearance is taken from his answers to a form sent to him from Chelsea Hospital in 1864.

'Col. Denis Packe commanded the regiment for a great many years, and his name would therefore be well known in it.

'Lieut. Anderson, the adjutant, did not enter the service until 1808; was adjutant from 1811 until after 1813, and therefore was the adjutant when *Michael* Gearyn was in the regiment.

'Michael Gearyn stated his age at enlistment into the 71st Foot in 1813 as 25. If Michael and Thomas were one and the same person, his age at death, October 28, 1871, would be about 83, not 105.

'The following extracts are intended to show the numerous contradictions that are in the book entitled " Longevity: The Life of Thomas Geeran, late of the 71st Highlanders."

'We give, first, statements made by the man himself, or by some one acting on his behalf, and then the extracts from the same work contradicting these statements.

'Appended to these are also extracts from the various letters and papers sent up to Chelsea Hospital from time to time in support of his petition for a pension for his services in the 71st regiment; and also evidence as regards the stations of that regiment during the period Geeran stated he served in it; its foreign service and history; nearly the whole of which tend to show that the 71st was not at the places at the time stated by Geeran, and that he could not possibly have served with it, and yet have performed the service he stated he did.

' This latter evidence is extracted from the " Historical Records of the 71st Highland Light Infantry," published by command of H. M. William IV. Compiled from official records by R. Cannon, Esq., Principal Clerk of the Adjutant-General's Office.

' The extracts from the book " Longevity " are printed in roman type, each extract being followed by its contradictory statement, *in Italics*, some of these being from the book " Longevity " and some from official records.'

' Pages 37 and 59. " Tom's father was a farmer, Tom assisted him. After his father's death he held the appointment of clerk in the office of a wealthy firm in Waterford, and was raised to an advanced post as agent to the branch house in America. Before starting he got drunk and enlisted."

' Page 56. *" Bred a sawyer, he continued to work at his trade till the year 1796, when he enlisted into the 71st Foot."*

' Page 39. " Sailed to join the 71st or *Glasgow* regiment in 1797.

' *" In June 1808, H. M. George III. was pleased to approve of the 71st bearing the title of* Glasgow *regiment."* (Vide *Historical Records.*)

' Page 39. " In 1797 they landed at Madras, where the recruits first met their comrades."

' " Seringapatam was taken May 4, 1799. Tippoo Saib was killed. Thomas Geeran did not see Tippoo killed, but *saw him* after his fall, and described him as a ' tall fine-looking fellow.' While this was going on the 71st were plundering."

'"*In October* 1797 *the regiment embarked at Madras for England. They were at sea during the remainder of the year, and arrived at Woolwich August* 12, 1798."

'"*During the year* 1799 *the regiment was stationed in Scotland. The head quarters were at Stirling.*" (Vide *Historical Records.*)

'Page 41. "Geeran said in the year 1801 the 71st was ordered to Egypt, and on March 21 at midnight Tom and his comrades were out and ready for battle."

'The late Marquis of Westmeath, on reading the above passage, denied it by saying "The 71st were not in Egypt at all." Geeran in reply said "My company was sent from Gibraltar, and I arrived at Alexandria with *Sir Denis Packe, General* in the Field and *Colonel* in the Army."

'"*Early in the year* 1801 *the* 71st *were in Dublin.* (*Left Scotland in June,* 1800.)

'"*On* April 24, 1801, *Lieut.-Col. Packe joined and assumed the command of the regiment.*

'"*The regiment remained in* Ireland until June, 1805.

'"*Major Packe was stationed with the 4th Dn. Gds. in England and Scotland until* 1800, *when he was promoted on* December 6, 1800, *to the rank of Lieut.-Col. in the* 71st *Regt., and on April* 24, 1801, *joined that corps in Ireland, in which country he served until August* 1805." (Vide *Historical Records.*)

' Page 42. Geeran's account of wound at Vittoria. Done by a Spanish soldier.

' *The Spaniards were the allies of the British, not the enemies, as asserted in the account of this wound.*

' Page 43. " Sir Thomas Picton, who commanded the ' 3rd Division, &c."

' " *Sir Thomas Picton commanded the 5th division at Waterloo.*" (Vide *Historical Records*.)

' Page 42. " Geeran received a ball in the left knee at Corunna, besides another gunshot wound."

' Page 49. Stated he was wounded at Waterloo in 1815.

' Stated in 1868 that he received a bullet or two in the body at Waterloo.

' Page 55. " *He escaped through Waterloo, and entered Paris with the victorious army.*"

' Dr. Pickford in 1864 stated in a letter that Geeran told him that he was wounded in the back at Salamanca.

' *The 71st was not at Salamanca.*

' Page 47. " I was *not turned out of the service*, but discharged from the 71st in 1819."

' Page 58. " *The fact of his not having a pension was owing, as he admitted himself*, to his having been discharged from the service for misconduct."

' Page 45. " He was discharged in the Isle of Wight, *invalided.*"

' Page 49. " *In confidence Geeran told a friend, &c.* . . . .

he *was* not an invalid *when discharged, but he thinks he was* dismissed *the service.*"

'Page 49. "States that about twenty years ago he received a Peninsula medal."

'Stated in 1864 that he received medals for Corunna, Waterloo, Peninsula (eleven clasps), and others. All made away with for drink or lost. (Vide *Chelsea Records.*)

'Page 58. "*The same cause (his misconduct) will account for his not having the Waterloo medal, which, under the circumstances, became forfeited, and also deprived him of any claim to the Peninsula medal.*"

'"*His name cannot be traced on the medal roll of men entitled to the Peninsula or Waterloo medals*" (Vide *W. O. Letter with Chelsea Records*).

'Stated he received 2*l.* 12*s.* 9*d.* in prize-money.

'*Name not found on prize rolls.*

'Page 44. "Geeran married in Gibraltar when he was thirty-five."

'*As he stated that he was thirty years old when he was enlisted, this would bring the date of marriage in the year* 1801.

'*From* 1798 *to* 1805 *the* 71*st was not stationed outside Great Britain.*

'CHIEF DISCREPANCIES IN GEERAN'S STORY.

'He stated that he joined the 71st in 1796, went out to India, and was at Seringapatam May 4, 1799.

'*The* 71*st left India in October* 1797, *and arrived at*

Woolwich August 12, 1798. *From that time until* 1805 *the regiment was not out of Great Britain.*

' He stated that in 1801 he was in Egypt, and that he went out with Sir Denis Packe.

' *The* 71*st was not in Egypt at all. Sir D. Packe was not out of Great Britain from* 1800 *until* 1805.

' In August 1805, the 71st went to the Cape of Good Hope. From there the regiment sailed, April 1806, to Buenos Ayres. The whole were made prisoners, August 1806, released, and returned to England, December 1807.

' Now all this was important service, yet Geeran does not mention one word about it.

' Stated he received prize-money and medals.

' *Name not on prize list or medal rolls.*

' States he was wounded at Salamanca.

' *The* 71*st was not at Salamanca.*

' States in one place he received a bullet or two in the body at Waterloo.

' *States in another part he escaped through Waterloo.*'

After this, few will doubt that the old fellow who so long traded upon the benevolence of the good people of Brighton, and on the credulity of some who ought to have known better was not Thomas Geeran, but Michael Gearyn ; and, so far from being 105, was only about 83 when he died.

The reader may think that some of my remarks on this case exhibit more warmth than the circumstances call for. Possibly it may be so. But I am an old public servant, and as such am naturally indignant at the want of candour and courtesy which has been exhibited towards the authorities of Chelsea Hospital ; and this, too, for the sake of bolstering up the impudent false-hoods of a gross impostor—for a grosser impostor, in my opinion, than the old man Geeran, or Gearyn, who called himself 105, but really was not 85, never existed.

JOHN PRATT, *not* 106.

The Report of the Registrar-General of Births, Deaths, and Marriages in England, for 1862, records the death in the Oxfordshire district of a man, aged 106 years.

This man was John Pratt ; whose name was first brought before me in a letter from the late Sir George Lewis, published in 'Notes and Queries,' of April 12, 1862 (3rd S. i. 281), which, containing as it does what that accomplished scholar really did say and think upon the subject of 'Centenarianism,' well deserves to be printed at length.

'It may, I believe, be stated as a fact that (limiting ourselves to the time since the Christian era), no person of royal or noble rank mentioned in history, whose birth was recorded at the time of its occurrence, reached the age of 100 years. I am not aware that the modern peerage and baronetage books contain any such case, resting upon authentic evidence. I have been informed that no well-established case of a life exceeding 100 years has occurred in the experience of companies for

the insurance of lives.   These facts raise a presumption
that human life, under its existing conditions, is never
prolonged beyond a hundred years.

'Nevertheless, the obituaries of modern newspapers
contain, from time to time, the deaths of persons who
are alleged to have outlived this age.   It may be con-
jectured that these statements of longevity are in
general made on the authority of the individual's own
memory.   Now, there are many reasons why old persons should be mistaken about their age, if their memory is not corrected by written documents.   Even
with persons in easy circumstances, great age is a
subject of curiosity, wonder, and solicitude; with
persons in a humbler rank of life, it is a ground of
sympathy, interest, and charity.   It is therefore not
unnatural that a person, whose real age exceeds ninety
years, and who has no contemporaries to check his
statements, should, without intending to commit any
deliberate deceit, represent his age as greater than the
reality.

'The only conclusive proof of a person's age is a
contemporary record of his birth, or the declaration of
a person who remembers its occurrence.   If there are
now persons living whose age exceeds 100 years, such
evidence surely can be obtained, and its production
would remove all doubt on the question.

'The writer of these remarks has investigated several
cases in which life was alleged to have lasted beyond
100 years, but it is difficult to obtain documentary
evidence of the fact.   The following case affords an

illustration of the result of such researches. A pamphlet has recently been published at Oxford by Mr.
Tyerman, a medical practitioner of that city, entitled
" Notices of the Life of John Pratt, now in his 106th
Year." In this pamphlet it is stated that John Pratt is
resident at Oxford, and that the writer of it is personally acquainted with him. The account of John
Pratt's birth and age given in it must therefore be
presumed to rest on his own testimony. The account
(p. 4) is, that " He was born at Grendon-under-Wood
in Buckinghamshire, on the fifth day of March, 1756,
and was the eldest of three children ; that his father,
who was a shoemaker, and a diligent man, died at the
age of 75 ; that his mother completed her 105th year,
and his great-grandmother her 111th." Through the
kindness of a friend, I have ascertained from the Rev.
M. Marshall, the incumbent of Grendon-under-Wood,
in Buckinghamshire, that the parish register of the
period (which is preserved) contains no entry of the
baptism of John Pratt at or near the year 1756,
although it contains various entries of baptisms, marriages, and burials of persons named Pratt from 1742
to 1783. The old man himself has no entry in a bible,
or other documentary evidence, in confirmation of his
statement ; and his account of his age appears to rest
exclusively upon his own memory.

' It is argued in favour of the belief in rare cases of
excessive longevity, that they would be in analogy with
other ascertained peculiarities of human physiology.
There have been men of extraordinary height ; there

have been minute dwarfs; there have been men of enormous fatness; there have also been men of extreme tenuity. Why then, it is asked, should there not be a few centenarians? This question may be answered by saying that such a duration of life does not seem, *à priori*, inconsistent with the laws of nature; but that the existence of very tall and very short, of very fat and very thin men, is proved by the indubitable evidence of eye-witnesses, whereas there is not on record, in published books, any conclusive proof of a life which has been prolonged beyond 100 years, under the existing conditions of our physical nature.

'I have, however, recently obtained the particulars of a life exceeding 100 years, which appear to be perfectly authentic, and to admit of no doubt. Mrs. Esther Strike was buried in the parish of Cranburne St. Peters, in the county of Berks, on the 22nd of February, 1862; she was the daughter of George and Ann Jackman; and she was privately baptized on June 3, and publicly baptized on June 26, 1759, in the parish of Winkfield, in the same county. She was therefore in her 103rd year. Certified extracts of the two registers proving these facts have been furnished to me through the kindness of the Rev. C. J. Elliott, Vicar of Winkfield.

'G. C. LEWIS.'

I then procured a copy of this 'Life of Pratt.' It consists of between forty and fifty pages; and when I say that, with the exception of a statement (at p. 28) that he married at twenty-three one Maria Dellamore, by whom (who died twenty-five years after) he had seventeen children ; (at p. 29), that he married again when eighty, there is not a date, place, or incident clearly or plainly stated to be found in the book beyond what Sir George Lewis has quoted,—the reader will probably share the feeling which its perusal excited in me—namely, that Mr. Tyerman was no more a match for John Pratt than good, honest, truthful Samuel Johnson for that arch-impostor, Richard Savage.

I am free to confess that, after a perusal of Mr. Tyerman's pamphlet, I was not surprised that the search for Pratt's baptismal register had proved fruitless ; and when I saw how full Pratt's story was of matter of no importance, and how entirely wanting in all that was required to establish its truth, I felt that if John Pratt was hardly one to whom could be applied the pretty saying, 'If he was not the rose, he had lived under its shadow ;' yet, when we consider how reticent he was as to the dates and events of his family history, and with what minuteness he relates the story of his being bewitched and going to a wise man for relief, and what the wise man said about Neptunus, Sol, and Luna, and which fills five or six pages in his biography, one is insensibly led to the belief that, if Pratt were not a gipsy, he had associated much with the tribe, and learned so much of their cunning, jargon, and power of mystification as to become quite an adept. The conclusion, in short, at which I arrived was, that Pratt's absurd figments were not only undeserving of credit, but were not worth the trouble of investigating.

Others thought differently. Among these is a lady of great experience and judgment in historical inquiries, and who, in 'Notes and Queries,' of 17th May following (3rd S. i. 399), expressed her confidence in the truthfulness of the old fellow, and gave the following account of him :—

'Having been personally acquainted with "Old Jack Pratt," during a residence of some years in Oxford, I must ask permission to record my firm belief that he is not a man likely to misrepresent his age for the sake of attracting sympathy. He is still living in great poverty ; and the following details have been procured from himself. My informant "found him much weaker, and," in her opinion, "he cannot live long."

'Old Pratt states that a copy of the register of his birth is in the possession of Miss D. Plumtre, of University College. (I have been told, not by Pratt, that Dr. Acland also has a copy.) He was not born in 1756, as stated in Mr. Tyerman's pamphlet, but in March 1755 ; this date he has always named both to my correspondent and myself. His eldest son, William Pratt, was born at South Shields, Northumberland (I think about 1783–8), and died in Shoreditch parish, at the age of eighty. Will any of your correspondents in these parishes verify these statements by consulting the registers ? The date which I have given above for William Pratt's birth, is not his father's statement, but my own deduction from some of his remarks, and may, therefore, be one or two years in error. I have not the honour of Miss Plumtre's acquaintance, but I would have ventured to ask her for a copy of the register had she been at home, which I understand she is not.'

This lady, in the ' Notes and Queries ' of the following week, inserted a slight correction ; but which, as it will be seen, contains nothing more definite as to Pratt's history.

' Though the Editor's note appears to close this subject, so far as persons under 120 years are concerned, I hope I may be permitted to correct a mistake in my former communication. Miss Plumtre *does not possess* Pratt's register (which cannot be found), but she has those of two of the brothers. The old man's memory has probably failed him in this matter ; he cannot remember the date of his eldest son's birth. He maintains, however, that he perfectly recollects the coronation of George III. in 1762. My correspondent adds that ' the doctors who have attended him say that the complaints from which he suffers are not those of a man of eighty or ninety, but of a much greater age. There are persons in the village where he was born who can recollect the family.'

My friend, the Rev. W. D. Macray—who obviously at that time shared the opinion of Pratt's truthfulness entertained by the lady just referred to—communicated to the same journal of June 7, on account of a visit made to Pratt on May 3 ; of which the following is the most important passage—indeed, the only one which treats of the question of his real age.

'With regard to his age, he gave as the date of his birth the same which is mentioned in Mr. Tyerman's pamphlet—viz. March 5, 1756, not one year earlier, as stated by your correspondent Hermentrude. With reference to the fact that the entry of his baptism is not found in the register of Grendon-under-Wood, he says that he was baptized privately when one week old; and, since registers were not kept with scrupulous exactness in the last century, as well as somewhat later, it is probable that the entry may through this cause have been forgotten. He states that he had a family Bible, in which the date of his father's birth, as well as of his own, was entered; that it was from this entry that his knowledge of the date was derived, and that he is certain of the accuracy of his recollection. This Bible he used to carry with him in his wanderings, until it was worn out; he then copied the entries on a paper, which he carried with him in a tin box; but at length, during one of his journeys, the box was lost, and with it was lost all the evidence he had of his age. I forgot to ask him where his first marriage took place, the register of which would, of course, afford sufficiently proximate proof concurrently with that of the baptism of his eldest son, as suggested by Hermentrude; but he incidentally mentioned, in the course of conversation, that the first of fourteen Scottish peregrinations was made in the year 1780, eighty-two years ago. It is hardly probable that a self-taught Oxfordshire "simpler,"— all of whose travels were made on foot—would be

M

induced to extend his tour to the wilds and moors of Scotland, for the sake of a few rare herbs not to be met with in the rich dells and woods of the South, before he had reached that age which, if Pratt's memory be correct, this year assigns.'

'Notes and Queries,' of September 6, 1862, contains the result of Mr. Macray's second visit to the old man :—

'I have made (as in my former communication I engaged to do) a few more inquiries respecting the age of John Pratt; and am now bound to confess, that either his recollection of events is remarkably and unusually treacherous, or that it is convenient to him that the events themselves which would prove his age should not be accurately reported.

'Towards the end of June I called on him again, and found him rapidly failing and confined to his bed. In the course of conversation, I asked him if he recollected where his first marriage took place. He replied, half laughing, 'I should think I do!' "Where was it?" said I. "At St. Martin's, Norwich." Upon going on to ask one or two more questions of date and name, he complained of pain and confusion in his head, and said he could not bear the attempting to think ; and so, having obtained the clue I wanted, I ceased to trouble him further. Through the kindness of Mr. J. M. Davenport of Oxford, and Mr. Kitson of Norwich, I am, however, now enabled to report that Pratt was not married at the church he mentioned. The latter gentlemen writes thus to the former :—

'"Search has been made at the churches of St. Martin-at-Palace and St. Martin-at-Oak (the only two St. Martins in this city), from 1770 to 1800; but no marriage of a John Pratt is to be found in either. In 1782, there is the marriage of *William* Pratt and Elizabeth Beck at the former parish."

'The fact that both baptismal and marriage registers are not to be found, creates a grave suspicion that Pratt's alleged age does not admit of proof; although his own appearance certainly shows that he has long passed the usual limits of man's longest life. It is observable that in Mr. Tyerman's account of him, his first wife (to whom he was married when he was twenty-three years old) is said to have borne the somewhat romantic name of Maria *Dellamore.* I am informed that the town clerk of Oxford (Mr. G. Hester) has been also making inquiries upon this subject with a view to publication, and that he does not give credit to Pratt's alleged age.'

It is pretty obvious from this that Mr. Macray— mildly as he expresses it—had lost all faith in Pratt, and had formed pretty much the same estimate of him as I had formed from the first.

As the late Sir George Lewis was the first to bring Pratt's case under the notice of the readers of ' Notes and Queries,' so to him were they indebted for sending to that journal of October 18, 1862, the notice of the old fellow's death which had appeared in the newspapers.

' Died, at the patriarchal age of 106, at Oxford, Mr. John Pratt, a native of Grendon-under-Wood, near Bicester. Deceased upwards of half a century ago, was

for many years employed in the herbal department of
Apothecaries' Hall, London, and was latterly well
known in Oxford, and many other parts of the country,
as a gatherer of herbs for medicinal purposes. He
retained his faculties in an extraordinary manner.
Shortly before his death he was seen enjoying his
walks through the streets of Oxford.'

And thus ends the story of John Pratt, alleged to
be 106 years of age, without one particle of evidence
in support of such allegation.

GEORGE FLETCHER, *not* 108 *but* 92.

My attention was not called to the case of Fletcher
until some years after his death, when among some
portraits of Centenarians which I purchased was one
from the 'Illustrated London News,' of March 10,
1855,—'The late Rev. G. Fletcher, aged 104. From
a photograph by Beard,' accompanied by the following
account of him. 'Mr. Fletcher was born on February 2,
1747, at Clarborough, in Nottinghamshire. From six
years of age he had been brought up in the tenets of
Wesley, and remained a member of that body till his
death. He spent eighty-three years of his life in active
pursuits. He was twenty-one years a farmer ; twenty-
six years he served his sovereign in the army—was
at the battle of Bunker's Hill, and followed Aber-
crombie into Egypt, where he gained the respect and
esteem of his officers. He then entered the West India
Dock Company's service, where he continued thirty-
six years, when he retired on their bounty, still pre-

serving up to within six months of his decease that astonishing activity of mind and body for which he was so remarkable; often travelling great distances by rail, and pursuing his holy calling, preaching two or three times a day, regardless of personal inconvenience, for the objects of charity and benevolence.

'The accompanying .portrait of Mr. Fletcher was taken on his 104th birthday, four years since. He walked for this purpose from Poplar to Messrs. Beard's photographic establishment, in King William Street, City; and after the sitting he walked back to Poplar, refusing to ride, although a conveyance was placed at his service.'

On reading this I recollected that I had seen—but unfortunately not 'made a note of'—a full exposure of the errors in this case, which I had vainly endeavoured to recover. I therefore addressed the following appeal to the readers of 'Notes and Queries' begging their assistance.

'In the "Times," of January 5, 1865, appeared a very admirable letter on Longevity. In the course of his communication the writer, who signed himself "A Pilgrim," after the very sensible remark that "some old people can recount events said to have occurred in early life, but when tested they are evidently circumstances which they have heard their parents relate," the writer proceeds to furnish an instance of this :—

' "The late Mr. Fletcher would occasionally preach in the Primitive Chapel in Nottingham, and the announcement that a man over a hundred years old would occupy the pulpit never failed to attract a numerous congregation. In his discourses Mr. Fletcher seldom failed to relate the sad events of a battle in which he had greatly distinguished himself in early life ; but at his death his friends procured his registry, which proved that he was not born when the battle was fought, and the particulars of the case were printed in 'The Nottingham Journal' at the time."

'I have long been very anxious to see the particulars of this case, of which I had heard. I have failed to obtain a sight of the journal in which it appeared, in spite of the courteous assistance of the gentleman by whom it is now conducted. Can any reader of "N. & Q." help me?'

This called forth only one reply—from Mr. John Bullock, which appeared in the same journal of November 25, 1871, and was certainly not calculated to impress one too favourably with the reverend gentleman's accuracy or disinterestedness.

'In reply to your correspondent, Mr. W. J. Thoms, respecting George Fletcher, permit me to state that I saw him stand and preach (in his way) for nearly two hours at Finsbury Chapel, Moorfields, on Wednesday, June 21, 1854. It was announced that he would preach two sermons on that day. Whether he preached the evening sermon, I cannot tell. I heard him in the afternoon of the above day give a sort of *vivâ voce* autobiography of his own life. The following is a correct copy of a bill in my possession relating to Fletcher's sermons, circulated rather freely at the time :—

'"*Finsbury Chapel, Moorfields.*—Two Sermons will be delivered Wednesday, June 21, 1854. Services to commence in the Afternoon at 3, Evening at 7, by the Venerable GEO. FLETCHER, in his 108th Year. For the benefit of an aged Minister."

'Portraits of the old man were sold in the vestry after the service, taken when he had attained his 106th year, viz. February 2, 1853 ; and stating that he had lived in the reigns of four kings, and her present Majesty Queen Victoria.'

In the meantime, having put myself in communication with the Rev. James W. K. Disney, the vicar of Clarborough, he kindly informed me that he had searched the Baptismal Register carefully from 1746, and could not find the name of George Fletcher till 1764, when the following entry occurs : '15 October, George, Son of Joseph Fletcher,' according to which, supposing no great delay to have taken place in baptizing him, he would have been about 91, and not 108 at the time of his death.

But there was another source of information available to me—the records of Chelsea Hospital, of which he was a pensioner ; and the kindness and courtesy with which General Hutt has at all times assisted my inquiries, and which I gladly avail myself of this opportunity to acknowledge with the publicity which they deserve, encouraged me to apply to him on this occasion ; and on so doing I was immediately furnished with the following particulars of Fletcher's age, and service in the army, from the records of that department.

'*3rd Foot Guards.*—George Fletcher, born in the parish of Clarbruf, in or near Redford, co. Nottingham, by trade a labourer; pensioned from Chelsea Hospital on April 18, 1803, at 1*s.* 2½*d.* a day, at the age of forty-nine years, died February 2, 1855, in 1st East London District.   Entered October, 1778.

'Service, $10\frac{6}{12}$ years in 3rd Foot Guards, and 14 years previously in 23rd Foot—total $24\frac{6}{12}$ years.

'Age returned by Staff-Officer of Pensioners at time of death, 108 years.   Age according to official records, 101 years.'

Anxious to ascertain the real age of Fletcher, General Hutt had search made in the Regimental Records on the possibility of their furnishing some particulars, and obligingly communicated the result to me.

From this it appeared that George Fletcher enlisted into the 23rd Foot, or Royal Welsh Fusileers, on November 2, 1785 ; and *deserted* on March 16, 1792.

He then enlisted into the 3rd Foot Guards, on March 14, 1793, and was discharged on April 20, 1803.

This information led to the hitting of blot number one in Fletcher's series of falsehoods.   His Bunker's Hill story was disproved.   To say nothing of the improbability of a boy, born in 1764, being present at an engagement in July, 1775, when only eleven years old, Fletcher did not enlist in the 23rd until 1785, ten years after Bunker's Hill.

But an official eye detected a yet more important blot.   Fletcher, after serving not quite seven years in the 23rd Foot, had *deserted*, and thereby forfeited his right to have his service in that regiment reckoned in his claim for pension ; whereas in his discharge he is not only credited with the seven years he had served, but with an additional seven years which he certainly had not served—altogether, fourteen years in the 23rd and ten years and a half in the 3rd Foot Guards, and received a pension of 1*s.* 2½*d.* per diem.

This led to further inquiries. The regimental books of the 23rd were again examined, and the accuracy of the original report confirmed.

The Description Books of the 3rd Foot Guards were then examined ; and from these it appeared that George Fletcher enlisted into that regiment about March, 1793. In these books he is credited with service in the 23rd Foot—*fourteen years !* No authority for this statement could be found ; the Letter Books for that period not being forthcoming.

Thus we see that this addition of seven years fraudulently added to his services in the 23rd, was managed somehow or other at the time of Fletcher's enlistment into the Guards.

That Fletcher was a party to the deceit can scarcely be doubted, for he must have known whether he had served seven or fourteen years.

It appears that he was entitled to reckon the seven years he actually served under a Royal Proclamation, issued in February, 1793, to the effect that 'all men who were then deserters from the British army, would be pardoned and their service restored, provided they gave themselves up to the various regiments stationed in Great Britain before a certain date, viz. May 1793.'

It remains to be added that Fletcher's age, when he joined the Guards, in 1793, was stated to be twenty-seven—which would have made him, when discharged in 1803, thirty-seven only—whereas, he is said in his discharge to be forty-nine!

But the reader may say, Do you mean to infer that in 1803 Fletcher added twelve years to his age, in view of his some day claiming the honour of being a Centenarian?

No: I mean nothing of the sort. Fletcher was shrewd enough to know that a man of thirty-seven years of age could not receive credit for twenty-four and six months' service, as much of that must have been before he was eighteen, and was consequently not entitled to reckon for pension. So in the same way that his seven years' service in the 23rd was transmuted into fourteen, his thirty-seven years of age was extended to forty-nine.

The twelve years added to Fletcher's age in 1803 stuck to him to the end of his days; and I have no doubt that by their deduction from the 104 years, which he claimed and was credited with, we shall have the real age of George Fletcher, namely, somewhere about NINETY-TWO!

George Smith, *not* 105, *but* 95.

I am indebted to the courtesy of a gentleman, a perfect stranger, for the following notice, extracted from a local journal :—

'REMARKABLE LONGEVITY.—On Sunday last (Nov. 6th, 1870), died at Ashstead Common, Surrey, Mr. J. F. Smith, at the extraordinary age of 105 years. Deceased was a very hale and robust man.'

My informant having added that he had left a widow who was herself 100, I felt interested in the history of a couple who had not only 'climbed the hill together,' but had tottered down it hand in hand for so many years.

The executor of Mr. Smith and the medical gentleman who attended him in his last illness were most courteous in replying to my inquiries, but were not in possession of any information calculated to show what his age really was. The latter, indeed, had no doubt that he was of the age stated, but his conviction was not based upon any evidence, merely upon what the old man believed he recollected.

But my inquiries having been brought under the notice of Mr. Smith's grandson, a most respectable tradesman in Westminster, he kindly put himself in communication with me ; and thanks to the information with which he furnished me, I was enabled to ascertain his real age.

In the first place, Mr. Smith informed me that his grandfather's widow, who survived till February, 1871, died then at the age, not of 100, but of 84.

In the next place that his grandfather's Christian name was *George*, and not *J. F.*, and that he was born at Merrow, near Clendon, as it was believed, in May, 1766.

A search in the Clendon Register proved unsuccessful. The Rector of Merrow was then applied to. He was from home at the time; he knew the case, but did not believe Mr. Smith to have been so old as represented, and promised to make search for his baptismal register on returning to Merrow.

He did so, and the result justified his and my incredulity, by showing that he was baptized on the 20th May 1775, making him 95, and not 105, at the time of his decease.

I ought to add that Mr. Smith's age at death is officially registered as 104.

EDWARD COUCH, *not* 110, *but* 95.

The 'Western Daily News' of January 31, 1871, announced the death on the preceding day, at Torpoint, of Edward Couch, who was 110 years of age. 'The deceased,' it went on to say, 'was one of the crew of the Victory at the Battle of Trafalgar, and was also present with Lord Howe at "the glorious 1st of June," in 1794. During his latter years he enjoyed a pension from Government, and his memory continued good until his death, which occurred yesterday.'

This confident statement provoked a controversy in the columns of the 'Western Daily News,' for the following analysis of which I am indebted to a friendly correspondent.

It commenced on the 1st of February, with a somewhat detailed account of his life, 'from a correspondent,' from which it appeared that Couch was born at Torpoint, in the parish of Antony, Cornwall, on Aug. 2, 1761; joined a privateer in 1780; was pressed on board the

'Romley' ship-of-war in 1790; fought on board the
'Gibraltar' under Lord Howe on June 1, 1794; was
taken by the French in 1797, and remained a prisoner
for two years and eight months; fought on board the
'Majestic' at the Nile, on board the 'Victory' at Trafalgar, and on board the 'Romley' at Boulogne; left the
service in 1816; and died at his native place, Torpoint,
on Jan. 30, 1871.

Feb. 3.   An account of his interment, with full naval
honours, in the presence of many thousands of persons,
in the churchyard of Antony.

Feb. 4. A letter from Mr. W. H. Pole Carew, of Antony,
stating that some 10 years since he examined the register of the parish (Antony) in which Couch was born, and
found that he was baptized in October, 1776, not in
1761; that the old man had been made acquainted with
the fact, but did not attempt to explain the apparent
discrepancy.

Feb. 7. Notice of a letter from 'A Relation,' who,
writing in reference to Mr. Carew's letter, said that a
younger brother of the veteran died at Torpoint in 1843,
at the age of 70, and would consequently have been 98
had he been living now; and that the parish register of
Antony would furnish the proofs.

Feb. 9. Notice of a letter from 'E. C.,' who contended
that the date of Couch's baptism was no proof of his
age, as his own (E. C.'s) father was baptized when four
years of age; and he (E. C.) knew a family of seven
children, all of whom walked, on the same day, three
miles to church to be christened.

Feb. 10. Notice of a letter from 'Torpoint,' confirming the statement of E. C. respecting Couch's younger brother, adding that there was a sister between them, and that the older brother was old enough to take care of the younger when both the parents went out to work ; and mentioning facts calculated to inspire confidence in the veracity of the old man, and in the correctness of his memory.

I am not aware whether this correspondence was carried on any longer in the local journal ; but Couch's case having been mentioned in 'Notes and Queries,' accompanied by a wish that it should be investigated, Mr. Carew, the gentleman already mentioned, forwarded the following communication, which appeared in that journal on the 4th March, 1871, (4th S. vii. 200) :—

' In reference to Edward Couch, whose name appears under the heading "Centenarianism" in "Notes and Queries," I addressed the following letter to the editor of the "Western Morning News" :—

' " THE LATE MR. E. COUCH.

' " Sir,— My attention has been called to a biographical sketch of the late Edward Couch in your paper of the 1st, in which it is stated that he was born in 1761.

' " Some ten years since the clergyman of the parish in which he was then living told me that this old man stated his age at that time to be near 100 years. He asked me to examine the register of this parish to ascertain the truth, and furnished me with the names of his parents.

'"I did examine the register, and found that he was baptized in October, 1776, not in 1761. The old man was made acquainted with the result of my search, but still persisted in his statement (and actually, some years later, referred to me as authority for its truth), though he did not attempt to explain his baptismal register appearing fifteen years later.

'"I leave it to you, Sir, and the public to decide whether, in sober truth, he died in his ninety-fifth or in his one hundred and tenth year. As these very exceptional cases of longevity are chronicled, I have thought it right to supply this evidence.

'"I am, Sir, your obedient servant,

'" W. H. POLE CAREW."

'"Antony, Torpoint, Devonport, February 3, 1870."

'Some of Edward Couch's friends, very loth to admit the possibility of his real age having been ninety-five instead of one hundred and ten, have argued that "he might have been baptized when he was fifteen, and that baptism in riper years is no uncommon occurrence." Another states that "his younger brother died in the year 1843, aged seventy years," and refers for proof of this brother's age to the register of this parish (Antony.) In reference to the first allegation it is at least singular that when told of the date as appearing in the register— as he was, to my knowledge, twice over—he did not say "I was fifteen when I was baptized." At that age such an event must have fixed itself in his memory. Moreover, I believe that baptism in riper years was at that period, the latter part of the last century, much more uncommon even than it is now. As to the second allegation, I have carefully searched the parish register, and cannot find this brother's name at all. Your correspondent W. C. thinks that this case may be easily tested at the Admiralty. Edward Couch's story describes him as pressed into the Navy in 1793—this is not at all improbable. If he was baptized at the usual time after his birth, he would have been seventeen in 1793—doubtless having been, as he stated, serving in a privateer before.

'I do not imagine that in those days, when the seaports were swept by press-gang crews, any very accurate report was sent to the Admiralty of the ages of the fish which they had netted.

'W. H. POLE CAREW.

'Antony, Torpoint, Devonport.'

Though I confess that I had very little doubt as to the real age of Couch, I felt that in this, as in all similar cases, it was desirable that the real facts should be ascertained ; and I accordingly applied for such information of his age as could be ascertained from the records of the Admiralty. And what was the answer ? that according to the age which Couch gave, namely, 19 when entering the Royal Navy, *on the* 30*th June*, 1794 —(the reader will remember Couch claimed to have served with Lord Howe 'on the glorious 1*st* of June,')— on board H.M.S. 'Bienfaisant,' he would have been 95 in June, 1870. I am bound to add that he was not a

Trafalgar veteran, for his name is not to be found on the books of the 'Victory.'

So much for Edward Couch, who claimed to be 110, but was, as proved alike by his baptismal certificate and the Admiralty Records, in his 96th year.

WILLIAM WEBB, *not* 105, *but* 95.

The following paragraph appeared in 'The Rock' of February 3rd, 1871, with the impressive heading 'More than a Centenarian' :—

'There has recently been much discussion as to the actual attainment of the age of one hundred years by man, and particulars have been given of some who have reached that age. The following particulars, which may *be relied on*, as to one who has passed his centenary, may be interesting :—There is now living at Frome, in Somersetshire, a man who has survived *a hundred and five winters*, and who is healthy, and capable of telling his own history. He was *born*—as he himself states— *at Frome, on Christmas-day, or Christmas-eve*, 106 *years ago*. In his early days he learnt the trade of a shearman (a branch of cloth manufacturing,) but when about 25 years of age he enlisted in the Marines, and served eight years, being present in some of Lord Nelson's memorable engagements. After he had obtained his discharge he returned to Frome and married, and has worked as a labourer until old age prevented his continuing to do so. He now rises about nine in the morning, and retires about nine at night. His memory is very good, and his hair not so white as that of many men at sixty years of

N

age. A few days ago he was a party to, and executed, a deed of conveyance relating to some little property in which he had an interest, and perfectly understood the nature of the document presented for his signature. Three years ago, this old man (whose name is W. Webb) was in the habit of cultivating his own garden, and was then seen bringing a faggot from the garden to his house (a distance of sixty or seventy yards) upon his shoulders, the weight being little less than fifty pounds, and of a most unwieldy size. Altogether Webb is a most remarkable man, and the facts being so well authenticated by collateral evidence, he is an object of much interest. He attends regularly at Christ Church, preferring that to the elaborate ceremonies of the parish church. Messrs. Penny, the local booksellers, have published a photographic portrait of Webb, which is sold for his benefit.'

When this confident announcement was placed in my hands, like the Sacristan, I 'said not a word to indicate a doubt,' but,—and there the parallel ends,—instead of applying my thumb and fingers, after his irreverent fashion, I took up my pen and wrote to Messrs. Penny for a copy of the photograph, and any confirmatory particulars with which it was in their power to furnish me.

In due time I received the photograph of Webb, which certainly did not suggest the idea of a man who had outlived a century for five years ; and in the letter which accompanied it Messrs. Penny assured me that the certificate of his baptism appeared in the church register, 'that he is so well known in Frome that there can be no doubt as to the authenticity of his age,' and enclosed a leaf out of last year's pauper list. It is published every year, and the Board of Guardians have thoroughly investigated the particulars of his age, and find it correct as stated in their list. On the page of the list enclosed appears the name of William Webb, and his age 105.

This letter was very satisfactory evidence that the Board of Guardians and the good people of Frome believed Webb to be as old as stated, but it did not contain a shadow of proof that he was so.

I then put myself in communication with a clergyman resident in the neighbourhood of Frome, the Rev. Thomas Waters, Vicar of Maiden Bradley, whom I had long known, and requested his assistance in eliciting the truth. This was readily promised, but the reader has little idea of the amount of trouble, and the extent of correspondence which my friend's kindness entailed upon him.

At first everything seemed to favour the supposition that the old man had really attained the great age claimed for him. Then came a letter that the verdict must be given against me, for that his baptismal register had been found as follows :—

'William Webb, son of Samuel and Rachel, baptized February, 1767.'

Then came the news that a brother of Webb's had been found, and 'interviewed,' and he was sure that his brother 'William was only about 90.'

At my suggestion the brother was visited a second time, and on being asked what was the Christian name of his and William's mother, said it was Elizabeth, thus .

confirming what the old man had admitted, that he was the son of Samuel and Betsy.

On this information further search was made in the register, when it turned out that the old fellow was baptized on the 10th August, 1776, making him not 105, but 95.

But this is not all. Just at that time my friend discovered in the possession of Mr. Dennis, the National School Master of Frome, Webb's discharge certificate from the Marines, of which the following is an abridgment :--

'Certificate of discharge from the Colonel-Commandant of the Royal Marines, Plymouth Division.

'William Webb, late a private in the above corps, was discharged on June 18, 1802, after 5 years 6 months and 4 days' service, on account of being undersized.

'He was then aged $21\frac{6}{12}$ years, 5 feet 2 inches high, brown hair, grey eyes, and fresh complexion.

'This certificate was given on Nov. 15, 1866, William Webb having lost his original certificate.'

This would make Webb some years younger. Whether for any purpose he made himself so when he entered the service it is impossible to say ; if so, he more than showed a balance towards the close of his life.

JOHN DAWE, *alias* DAY, *not* 108 *to* 116, *but* 87.

In the summer of 1871 I received from a gentleman of great respectability, whose name it is needless to state, the following letter, in which my attention was called to the case of a man who was believed to have died at the age of 116, he having lived one hundred years in the family of the writer :—

'My dear Sir,—In a branch of my own family there lived and died an old servant who was one hundred years in the family.

'His name was John Day. He came as a parish apprentice, and died at the age of 108, having lived all that time, that is, four successive generations of John Rogers, Esq., of Holwood, in the parish of Quethiock, Cornwall. He was buried in that parish by the Rev. Dr. Fletcher, the present vicar.

'The family of Rogers have now left Holwood, but Charles Rogers, Esq., Newbury, Berkshire, is the representative. He and his sisters still live there, and they as well as myself knew the old man when we were children.

'He was reputed to have been 116, having been christened when he was eight years old.

'When he died, my father calculated by his daily average consumption, that he must have drunk enough cider to have floated a first-rate line of battle ship.

'Dr. Fletcher's father died a few years since, at 102, I think I have heard.

'Understanding that you are engaged on a work on Longevity I thought that these facts might interest you. I am sure that either Dr. Fletcher or Mr. Rogers would reply to any communication on the subject

'Believe me, yours truly,

'—————.'

Presuming from my correspondent that Dr. Fletcher was well acquainted with the case referred to, I addressed a short note to him, asking him for any particulars with which he could oblige me ; and was not a little surprised at receiving from the reverend gentleman the following courteous, but unsatisfactory, reply :—

'The Vicarage, Qnethiock, Liskeard : Aug. 5, 1871.

' Dear Sir,—I beg, in reply to your letter of the 4th inst., to say that I shall be very happy to make all the enquiries possible about the case of John Day. But, to help me in them, will you tell me all you know of the date of his death ? My register commences in the year 1574, and many entries at these early dates are faded and unintelligible. The labour of wading through 200 years of this register would be long and tedious, unless you can furnish me with some boundary of time.

' I have searched my register at different times with different objects, and have no recollection of having met with the name of John Day.

' I think that if such a name, with a note of any advanced age had occurred, I should have noted it. In the early portions of the book the entries are in Latin. but very correct as far as they can be deciphered.

' My late father (the Rev. John Kendall Fletcher, D.D., of Callingh, Cornwall, a Magistrate, Commissioner, &c., &c.,) died about twelve years ago, in his one hundredth year.

' On the floor of the church there are some fine brasses 400 years old, in commemoration of a family of the

Kingdons who had property here, but their ages are not
recorded.   It appears from the inscription that not only
the persons named here, but "omnes progenitores eorum,"
were buried in the church.   Many of the descendants of
this family now live in different parts of England.   In
Derbyshire, Devon, and Cornwall.   I take you to be an
antiquary, and interested in these things.

' I am, faithfully yours,

'JOHN R. FLETCHER.

'W. J. Thoms, Esq.'

Of course I lost no time in furnishing the worthy
Vicar of Quethiock with all the information contained in
my correspondent's letter, and in due course received the
following obliging and satisfactory reply :—

' The Vicarage, Quethiock, Liskeard : Aug. 11, 1871.

'Dear Sir,—I find that the case you write about
requires much correction.

'I suspected from your first communication that the
name and circumstances were incorrectly set forth.

'From my long acquaintance with our registers, and
frequent careful inspection of them, I had strong doubts
about the name " Day " ever occurring in them, and this
was verified by inspection.   But now that my memory is
assisted by your recent communication, I am able to set
you right.

'I well remember the old servant of Mr. Rogers, at
Holwood.   He was kindly, in his old age, located in the
chimney corner of the hall there.   I buried him at Quethiock on the 19th of August, 1828, and his age (alas for

the unreliable result of information dependent on popular stories), his age is recorded as 87 years.

'I have unsuccessfully sought for the register of his baptism. There is an hiatus in the register about the probable period of his birth, nor does it follow that he was a native of Quethiock.

'Instead of John "Day" his name was John "Dawe." I remember the old man well, and I believe his age was properly recorded.

'One man was buried here September 23, 1863, whose age is recorded at 99 years, which I have reasons for believing to be correct ; his name George Snell, a respectable farmer, who to a period not long preceding his death was able to move about with a large measure of facility.

'I am, dear Sir, yours faithfully,

'JOHN R. FLETCHER.

'W. J. Thoms, Esq.

'I enclose with this the "Times" extract.'

This case strikingly illustrates how comparatively worthless in matters of this kind is that species of evidence which is too frequently regarded as beyond dispute—namely, 'personal knowledge.' John Day, of 108 or 116 years of age, according to popular belief, proving to be only old John Dawe, of the less remarkable age of 87.

GEORGE BREWER, *not* 106, *but* 98.

In the 'Guardian' of the 20th September, 1871,
appeared the following paragraph, copied from the
'Hampshire Telegraph':—

'On Thursday week, *Robert* Brewer died in a court in
Havant Street, Portsea, at the remarkable age of 106 years
and one month. We learn that he was born at Gosport
on the 7th of August, 1765, and, though it might have
been supposed that the deceased had, as other veterans
have done, made a mistake as to the year, the exactness
of the date is confirmed by the events at which he was
present. When about *twenty years of age he joined the
navy*, where he remained until he was twenty-nine. He
was in an action off the coast of Ireland, was on board
the flag-ship of Admiral Byng during the Mutiny of the
Nore, and was captain of the maintop on board the
"Robust," one of the ships forming Lord Howe's fleet
on the glorious 1st of June, 1794. On the last occasion
he was severely wounded by a cannon shot in the left
ankle and side, and was discharged with a pension of
11*l.* per annum, which six years ago was augmented to
20*l.* 12*s.* Since that time he has lived in Gosport and
Portsmouth, and carried on the calling of a waterman and
fisherman, until about six years ago, when he was compelled, on account of the wound in his foot, to retire
from active work.'

As this seemed to be a case which might be tested
without much difficulty, I applied to a friend at Ports-

mouth, through whom I procured a photograph of the 'old salt,' taken when he was supposed to be 99, and heard that his name was not Robert but George, and that there was some difficulty as to procuring his baptismal certificate.

But there was still one resource left, the books of the Admiralty, and from these it appears that the statements that 'he was about twenty years of age when he joined the navy,' was so far correct that he gave that age when he joined on the 19th of March, 1793, which would give 1773, and not 1765, as the year of his birth; that he was not on board the 'Robust' on the glorious 1st of June, not having joined the 'Robust' until the 15th of February, 1795, and was discharged incurable on the 17th of May, 1799. So the old fellow had slipped on about eight years, as it is clear if he was twenty in 1793 he was not 106, but 98, when he died in 1871.

Robert Howlison, *said to be* 103.

'Robert Howlison, aged 103. The instances having been so frequently recorded in the public ·journals, and so minutely examined in "Notes and Queries," the place and date of each fresh occurrence ought to be forthwith laid before its board of inquiry.

'In last Monday's "Echo" (Jan. 23, 1871), I read the pleasant account of a purse of twenty-five sovereigns having been presented to Robert Howlison, of West-Linton, Peebleshire, on his *hundred and third* birthday. Most cordially do I, who am in humble expectancy of my *ninety-fourth*, wish my venerable senior "multos et felices," with the like testimony attached to every one of them.'

So wrote in 'Notes and Queries,' of February, 1871, a venerable and much respected correspondent; and to this communication was added a few words expressive of the editor's hope that some Peebleshire correspondent would kindly furnish the evidence of Robert Howlison's age.

The appeal was not attended with any result; but after the old man died, on the 30th of October, 1871, there appeared in the 'Peebleshire Advertiser,' of the 25th November, the following memoir of him, and I beg the reader to pay especial attention to the concluding paragraph :—

'On the 30th ultimo, there died at West-Linton, Peebleshire, Robert Howlison, who had almost reached the age of 103 years. A short notice of his death appeared in the columns of the " Peebleshire Advertiser" a fortnight ago ; but we think that not only on account of his extreme old age, but also from his being a type of Scotsmen now rarely to be met with, he deserves more than a passing obituary paragraph. The deceased was born at Channelkirk, on Handsel Monday, 1769. He had the misfortune to lose his father when only nine years old, which caused Robert and his elder brothers to be sent at an early age to work. He was engaged in country service, and for many years followed the occupation of a ploughman ; but falling into indifferent health, he exchanged his calling for that of a shepherd, which he pursued till upwards of eighty years old, when the infirmities of old age compelled him to

relinquish it. Though necessitated from his earliest
years to labour, he did not neglect the means either of
mental or spiritual improvement. He was always a good
reader, especially of those religious classics long so dear
to the pious of our land, the perusal of which tended, in
no slight degree, to gain for our countrymen the proud
distinction of being the most intellectual peasantry in
the world.

' Robert, in his early manhood, lived at Stow, and
while there joined the Secession Church, at that time
under the pastorate of the Rev. Mr. Kidston, father of the
late Dr. Kidston, of Glasgow. After applying to be
received into membership, he was kept back a year that
the minister might have his eye upon him during that
time, for as Robert said, "it was nae bairn's play to join
the kirk in thae days." About the beginning of the
present century he removed to the neighbourhood of
Peebles, where he attended the ministrations of the Rev.
Mr. Leckie, who was the first Secession minister there.
Afterwards he resided in Newlands parish, where he
long remained, and attended Linton meeting-house.
Here his becoming deportment, sagacity in counsel, and
well-stored mind, pointed him out as a fitting person
to perform the duties of the eldership, and he was
ordained, along with one still surviving, to that office.
His last place of service was in Dumfriesshire, and when
unfitted longer to pursue his daily toil he returned to
Linton, where he lived until his death. For many years
he described himself as waiting—and never did aged
saint, with more child-like faith, wait till his change

came, than did the subject of this brief memoir. Sustained by a wonderful buoyancy of spirit, he relished a good joke, was full of anecdotes and wise sayings, and could delight a company with his sallies of wit. He could recall the arguments of preachers he had heard in his youth, and combat their opinions when he thought them unsound; at the same time he could take a lively interest in events which were passing around him. He recollected incidents connected with the early life of the first Napoleon, when as yet the great events of modern European history were unperformed and unwritten, and to the last he could read books with pleasure and profit. But it was as a man of prayer that he specially excelled, and none who heard the fluency of his devotions but felt they were in the presence of a man of superior religious attainments.

'When fifty years of age he married, and his widow survives him. They have had a numerous family, several of whom are in distant lands. Although unfortunately towards the close of his life he lost the greater part of his savings, he never murmured, nor would think of accepting parochial relief, alleging that God, whom he had served for a life-time, would sustain him during the few years he had to live. To mark their sense of his worth, as also to help to soothe his declining years, a sub-scription was got up for him, a year ago, by many friends; and on his last birthday he was presented with the sum of 25*l.* along with a neatly-written address. He has not lived long to enjoy it, but has exchanged the poverty of this world for the riches of the everlasting kingdom.

'We saw, some time ago, that one had been enquiring in "Notes and Queries" for proofs of Robert Howlison's age. These are numerous enough, and their veracity cannot be questioned; but as it is not our purpose to mention them here, we simply refer the writer to his friends in West Linton.'

But if this notice, which the writer undertook because on account of his 'extreme old age,' among other things, Howlison deserved 'more than a passing obituary paragraph,' is singularly deficient in facts and dates, it was satisfactory to learn from it 'that *proofs of Howlison's age were numerous enough, and their veracity not to be questioned.*'

I *now* know why the writer did 'not mention them,' and why he 'simply referred the writer to his friends in West Linton.'

The proofs *do not exist.* Having no friends in West Linton, I wrote to the editor of the 'Peebles Advertiser,' stated I was the 'writer' alluded to, and expressed my hope that the proofs referred to would be produced, more especially as there was a discrepancy as to Howlison's age between the two accounts given of him. It being stated in January that the purse was presented to him on his 103rd birthday; whereas the later account spoke of him as having 'almost reached the age of 103 years.'

The editor, I presume, referred my letter to the writer of the article in question, who, having nothing to say, very wisely said nothing.

At all events, no answer ever reached me.

Some time afterwards Dr. Ramage, to whom I am indebted for many similar kindnesses, made inquiries for me into the case of Howlison, and in due time forwarded me the result of his inquiries.

The papers which Dr. Ramage placed in my hands furnished abundant testimony to what I had never thought of questioning, the excellent character of Robert Howlison ; the great respect in which he was held by all who knew him, and that he certainly was a very old man. The 'numerous proofs' of which 'the veracity could not be questioned,' were conspicuous by their absence.

What had satisfied Robert Howlison's friends that he was nearly 103 were, as the reader will see, *not proofs but inferences ;* and as Dr. Ramage well observes 'a year or two deducted from Howlison's reputed age of nine at his father's death would make all the difference.'

In the first place, no register of either the birth or baptism of Howlison has been found or produced.

He was the second eldest son of a miller, residing at Channelkirk, Berwickshire. At the age of *nine* he lost his father ; but it is neither proved that he was nine at that time, nor is it proved when the father died, although these two supposed facts are the chief foundation on which his age has been calculated.

Mr. Coutts, of Bell's School, Leith, writes :—'Ten years ago, in 1862, Robert Hunter, farmer, at Amozendean, Carlops, died at the age of 84. He told me he had made bands or straw ropes to Robert Howlison 74 years before. At that time he said Robert Howlison was a

man, and working a man's work at the scythe, and
would not be younger than 19 he thought.'

Mr. Coutts further states, that 'once a few years ago
a brother of Mr. Howlison's was on a visit to the old
couple at West Linton, and Robert, *anxious to know for
certain how old he was,* mentioned the name of a family
near by his father's, with whom they were on terms of
intimacy. A daughter of this family was born a month
after Robert. The brother-in-law being in the East
country a few weeks after, got the records looked up.
He *did not find* Robert's name; but found that of the
female, whose birth agreed with the time stated, viz.,
February or March, 1769. This he at once communicated to Robert, who knew that old Hansel Monday
was his birthday, but was uncertain as to the exact
year, which *he thought was* 1770. From that time the
question seemed, at least to the minds of himself and
family, entirely set at rest.

There is one other fact mentioned in corroboration,
namely, that his widow, to whom he was married upwards of 52 years, and who died shortly after him, at
the age of 77 years, used to say that he was 25 years
older than herself, she being just upon 25 when married,
and Howlison being 50.

Without 'questioning their veracity,' these cannot be
considered 'proofs enough to establish the fact' that
Robert Howlison had reached the exceptional age of
102 and 10 months; and I think I may fairly claim in
this case the sensible Scotch verdict– NOT PROVEN.

ROBERT BOWMAN, *not* 118 *or* 119.

I have upon more than one occasion been weak enough, in spite of my better judgment, to enter into the examination of a case, which I had little hope of being able to bring to a satisfactory conclusion ; and I have been induced to do this because I did not choose to submit to an insinuation very freely made, that I was afraid to go into it lest it should upset my supposed theory.

Such a case is that of Robert Bowman ; and the reader, when he becomes acquainted with it, will, I think, be struck at the confidence with which its truth is assumed by a man of professional eminence in the absence of anything worthy of the name of evidence. Bowman's case illustrates one of several canons which may safely be laid down in cases of alleged Longevity ; namely, that when the supposed Centenarian is believed to be a hundred, or a year or two over, some error may not unreasonably be *suspected* ; but when the age is extended beyond, say 106, error so certainly exists, that no trustworthy evidence can be produced in support of it.

The case was first brought forward by ' Notes & Queries,' of July 20, 1870, 4th S. vi. 91, by the Rev. Canon Harcourt, of Carlisle, in the following letter : —

' THE CENTENARIAN BOWMAN.

' I believe the case of Bowman, who died when upwards of one hundred and sixteen, has occurred in your paper, but has not elicited any remarks.

' Sir George Lewis was ready to admit cases of Cen-

O

tenarianism where there was a well evidenced register,
and I think you have only excused yourself in doing so
by supposing some elder brother of the same name.   In
Bowman's case there are three evidences, which the
world in general would deem satisfactory :—Dr. Barnes,
who published an account of him, and who was long the
principal physician in Carlisle ; Mr. Mounsey, an eminent solicitor at the same period ; and the incumbent of
the parish.   This, I think, would be enough for anybody
who believes in registers.

'There is, besides, the corrobation of the best possible
witnesses :—Mr. Mounsey, the son, also a solicitor ; Mr,
Saul, the chapter clerk ; W. N. Hodgson, Esq., M.P. ;
Mr. Graham, of Edmund Castle, the great-nephew and
heir of the person on whose estate Bowman was born,
and who was his landlord.   He worked at one time as
a labourer in the trenches at Carlisle in 1745 ; but had
acquired money enough to buy a house and small property near Edmund Castle, in which he died.   Mr.
Graham often visited him.   He used to bring his half-
yearly rent of 10*l.* in a stocking to Edmund Castle.   The
late Mr. Howard of Corby, his son the present proprietor,
the Bishop of Chester (Dr. Law), Lord Carlisle, and
various others well acquainted with the family, visited
him.   He had a son who was eighty when he (Bowman)
died, and another who was seventy-three.   If this case
is not to be believed in, it is impossible that any others
can.   I asked my surgeon, Mr. Page, whose practice
reaches from Liverpool to Glasgow, whether there was
any doubt about the matter.   He said there was not the

slightest.   His opinion, in my mind, far outweighs that
of all the amateur sciolists who think they know something of physiology.   But I think you are aware that
real physiologists are content to follow Dr. Harvey, as I
should think eminent and learned men might be to
follow the opinion of Lord Arundel.

' I have seen no reason whatever to doubt of the case
of Parr.   In the case of him and Jenkins, it can be of
no use to ask questions of persons in Shropshire and
Yorkshire.   There is a discrepancy in the case of Jenkins, which I have noted, of seven years; and what is
wanted is an accurate copy of a paper said to be preserved in the office of the Queen's Remembrancer, and
any other papers which may be there on the subject.

' Mr. Page says that a doubt about Centenarianism is
like a superstition of a London cowkeeper, who said he
had kept 999 cows, but that it was impossible to keep
1000.

' Mr. Page has conversed with a Centenarian of one
hundred and four, whose account of himself he considers
to be perfectly correct.

' The other day he was visiting a family, one of whom
died when within two months of one hundred, and he
knows numerous instances of Longevity.   His son visited
the other day at Richmond, in Yorkshire, a person who
was a Centenarian in April, and of whose case there is
no doubt.   There was no trace of the *arcus senilis* in his
eyes.

' C. G. V. HARCOURT.'

' Carlisle.'

To which was appended an Editorial note that Mr. Harcourt was mistaken in believing that Bowman's case had been discussed in ' N. & Q.,' and that the Editor was ready to receive the evidence to which he refers, but must remind him that, as the case is very exceptional, it can only be established by evidence which will bear the strictest scrutiny.

This was followed on August 18, by another letter from Canon Harcourt, of considerable length, naming many highly respectable persons who had seen Bowman, and did not doubt his age, promising to send a copy of Dr. Barnes' pamphlet, but containing no further evidence ; and on September 3, by a short note from Mr. Sidney Gilpin, E.C., of Carlisle, to the effect, that he was sorry to find that Mr. Harcourt had neither examined nor obtained a copy of the register at Hayton to verify Bowman's reputed age; and his willingness to do what he could towards examining the entry, and also to get a copy of it signed by the minister of the parish.

Having in the meanwhile received and read with signal disappointment Dr. Barnes's account of Bowman, I urged Mr. Gilpin, in ' N. & Q.' of September 10, to undertake the investigation, pointing out to that gentleman the missing link in Dr. Barnes's narrative, and venturing to suggest points to be looked to. The following is my letter :—

' If Mr. Sidney Gilpin will kindly investigate the case of Mr. Robert Bowman, who died at Irthington on June 18, 1823, as Dr. Barnes supposes, in the one hundred and eighteenth year of his age, he will be doing good service to the inquiry now going on with respect to human Longevity.

' Dr. Barnes' account of Bowman, full as it is of inte-resting physiological details and personal anecdotes, *does not contain one tittle of evidence* on the points on which the whole case rests, namely, the identity of the Robert Bowman baptized at Hayton in 1705 and the Robert Bowmen living at Irthington in 1820. Dr. Barnes, will, I trust, forgive me for entertaining a doubt upon this subject—a doubt which is strengthened by the fact, that whereas the supposed Centenarian " believed he was born about Christmas," the Hayton Bowman was not baptized till September or October.

' I think, if Mr. Gilpin searches the registers of Hayton or of Tottington, which it appears from Mr. Harcourt's letter is the adjoining parish, he will probably find the real register of Dr. Barnes's hero ; who will, I suspect, turn out to be the son of the Bowman baptized in 1705.

' Bowman, it appears, married when he was fifty, i. e. in 1755 ; but his eldest son was only fifty-nine in 1820 ; from which it would appear that, though Bowman had six sons, all of whom were living in 1820, the eldest was not born till five or six years after his marriage.

' Perhaps, if Mr. Gilpin could find the certificate of Bowman's marriage, it might throw light upon the ques-tion of his age and identity.'

The inquiries which Mr. Gilpin so liberally undertook satisfied him as to the exceptionally great age of Bow-man. I do not anticipate that any of my readers will share that gentleman's convictions ; but it is only justice to him that I should publish, at full length, the grounds on which he rested his belief, as furnished by him to ' Notes & Queries' of December 31, 1870.

' I have spared neither time nor trouble in investigating the case of Robert Bowman,[1] having personally examined the registers of four different parish churches, besides making numerous inquiries respecting the registers of more distant churches and chapels by letter or otherwise. I have likewise been in personal communication, with a grandson and granddaughter of Robert Bowman, and two grandsons of his younger brother Thomas,[2] all of whom were extremely anxious and willing to assist in bringing my labours to a successful termination. As might naturally be expected from the lapse of time which has transpired, the information obtainable at this date is not on all points so full or so many-sided as could be desired. Nevertheless there is ample evidence, I think, to convince any but such as are unduly burdened with sceptical minds on the subject, that Robert Bowman was what he represented himself to be—that is to say, he was at least *one hundred and eighteen years old* at the date of his death.

' In the first place, the Hayton parish register was gone through carefully for fifty or sixty years, and the only baptism bearing directly upon the subject is the

---

[1] If Mr. Gilpin, a resident in the neighbourhood, found the investigation of one case entailed so large an amount of time and trouble the reader may readily imagine how much of both the sifting of the several cases related in the present volume has cost the writer.

[2] Thomas Bowman died at Grinsdale, near Carlisle, in 1810, aged ninety-nine years, or as some assert, one hundred and three years.

one mentioned by Dr. Barnes, entered in the year 1705 (between September 23 and October 28), which, by being written at the bottom of the page, has left nothing clearly discernible, except the name and place of birth as follows: "Robert Bowman of Brigwoodfoot."[1]

'It is believed that Bowman lived in the neighbourhood of Corby Castle about the year 1755, the supposed time of his marriage. We therefore examined the Hayton, Irthington, Wetheral, and Warwick registers for his marriage certificate—running through each of them a good many years—but without success. In the register of burials at the Irthington parish church there occurs the following entry :—

"Robert Bowman, Irthington, June 23rd, 1823, aged 118 *years.*

"JOHN TOPPING, Vicar."

'A chaste stained-glass window has been inserted in Irthington church to the memory of Bowman by his youngest son, and in the churchyard he has also erected a massive tombstone bearing these inscriptions :—

"Robert Bowman, Yeoman, of Irthington, died 18th June, 1823, at the patriarchal age of 119 *years.*

"Elizabeth, his wife, died 22nd March, 1807, aged 81 years.

"John, the eldest son, died 29th July, 1844, aged 84 years.

[1] The Rev. George Toppin, the present incumbent of Hayton, writes: 'This entry being at the foot of the page, and much worn, I cannot ascertain distinctly the remainder of the entry, but I can see there has been a proper filling up.'

" Robert, the second son, died 19th Sept., 1825, aged 62 years.

"William, the fourth son, died 23 Dec., 1836, aged 68 years.

" Thomas, the fifth son, died 28th Sept., 1853, aged 83 years.

" Joseph, the youngest son, died 20th Nov., 1857, aged 84 years."

'The remembrance of incidents occurring in child-hood, or in early years, presents a marked feature in the memories of most aged people. An old man said to me the other day, in his own homely language : "Why, bless ye, I's gittin' quite doaty, an' forgit maist things 'at happen noo-a-days ; but weel I mind many a thing 'at happen'd lang syne when I was a bit boy." And so it was with Robert Bowman and his younger brother Thomas.

' Robert had a distinct recollection of witnessing the following incident connected with the rebellion of 1715 : —A guard belonging to the royal troops was placed on the bridge at Newby, in order to intercept the return of any rebels who might be making their way into Nor-thumberland. A Jacobite officer or horse soldier, called Fallowfield, on approaching the bridge, and seeing the danger he was exposed to, left the highway just as the king's troops opened fire on him, and galloped in hot haste through the fields until he came to the river Irthing, which he crossed in gallant style and so escaped.

' Thomas Bowman was a boy scarcely out of petticoats when the first rebellion broke out, and often used to relate that a party of soldiers with a baggage waggon cried out to him in derision, as he stood gazing with boyish wonder at their white cockades and gay colours : " Come, me lad, jump up ahint, an' show us t' nearest cut across t' country ! "

'Thomas, when young, worked for the ancestors of the present Sir Robert Brisco of Crofton Hall, near Carlisle, for a groat a-day. He afterwards settled on a farm in the neighbourhood, and, what is very remarkable, lived under the Brisco family as husbandman and farmer for *more than eighty years.*

'I will now proceed to state briefly the different points on which I rest my belief in the genuineness of Robert Bowman's great age.

'In the first place, I have faith in the simple, straightforward, and apparently truthful and consistent narrative related by Dr. Barnes, which, it must remembered, was made public *three* years before Bowman's death.[1]

'Secondly, after carefully searching the registers of four adjacent parishes, no entry of any kind has turned up to show that any person of the same Christian name and surname has been baptized at a later date, *i.e.* within a reasonable time.

'Thirdly, Bowman having passed his whole life in the neighbourhood of his birthplace—excepting a few early years spent in Northumberland—is in itself a significant fact, and one which destroys all ordinary chances of

---

[1] 'The first notice of Bowman as a centenarian was contributed by Thomas Sanderson to the " Carlisle Patriot," in 1817, six years before his death. As a natural consequence, Dr. Barnes goes over some of the same incidents, but is fuller in the different details and more concise.'—S. G.

flagrant deception ; such, for instance, as a man personating his own father or any other person whatever.

' Fourthly, if Robert Bowman's age be a delusion and a snare, then is also the age of his brother Thomas. The two men must stand or fall together.

' If we may believe some of the prophets who have prophesied, the county of Cumberland is remarkable above most counties for the longevity of its inhabitants. Joshua Milne, of the Sun Life Assurance Office, writing to Dr. Heysham of Carlisle in 1812, says :—

' " Being engaged in inquiries relative to human mortality, and having met with your valuable observations thereon, that were published at Carlisle in 1797, I have constructed a table of mortality from them, whereby it appears that the inhabitants of your city surpass in longevity those of any other place (so far as I am informed) for which a similar table has yet been constructed."

' In Lysons' " History of Cumberland " (p. xv-lii.) a list of no less than one hundred and forty-five individuals is given, who died between 1664 and 1814, aged from 100 to 114 years—one person being cited at 135 years. The Messrs. Lysons state that, in some cases, they had opportunities of ascertaining the accuracy of the ages.'

I do not know whether I was more disappointed because Mr. Gilpin's persevering researches had been attended with results so little commensurate with the labour they had cost ; or more surprised at the conclusions which Mr. Gilpin drew from them. To Mr. Gilpin, they seemed to establish the abnormal longevity of Bowman ; to me, on the other hand, not only to justify, but to confirm, the doubts which Dr. Barnes' account had awakened in me; and while acknowledging, in 'Notes and Queries,' January 10, 1871, (4th S. viii. 38) the good services which Mr. Gilpin had performed, I explained my reasons for differing from his conclusion, in the following reply :

'Mr. Gilpin deserves the best thanks of all who are interested in the question of Longevity for the trouble he has taken in investigating the case of Robert Bowman ; and as one who knows by painful experience the vast amount of time and labour which such inquiries entail, I beg to thank him most heartily.

'I appreciate the good service he has done in collecting the information which he has laid before the readers of " N. & Q.," and I am the more anxious to avow this, seeing that, at the risk of being classed among those "who are unduly burdened with sceptical minds on this subject," I am so far from drawing from the evidence brought forward by Mr. Gilpin the conclusion at which he has arrived—viz., that Robert Bowman was "at least 118 *years old* at the time of his death "—that my doubts upon that point are very considerably strengthened.

'So far from confirming or establishing the identity of the Robert Bowman, baptized at Hayton in the year 1705, with the Robert Bowman who died at Irthington in 1823, the evidence adduced by Mr. Gilpin seems to have a directly opposite tendency. Mr. Gilpin searched the Hayton register carefully for fifty or sixty years, and the only baptism bearing directly upon the subject is that of Robert Bowman, baptized in 1705 ; but if this is the baptism of the centenarian Robert, the same register would, in all probability, have contained the register of the brother Thomas, said to have been born either in 1707 or 1711. Surely the absence of the baptism of Thomas leads to the inference that the Robert baptized was not the brother of Thomas, and consequently not the Robert who died at Irthington. Mr. Gilpin, who produces not a tittle of evidence as to the age of Thomas, "who died in 1810, aged 99 years, or, as some say, 101," says : " If Robert Bowman's age be a delusion and a snare, then is also the age of his brother Thomas. Both men must stand or fall together." I agree with Mr. Gilpin in his premises, but differ in his conclusion. I hold that there is not a particle of evidence as to the real age of either of them.

' It is much to be regretted that Mr. Gilpin's endeavours to procure the marriage certificate were not attended with success ; as, although such certificate would probably not have shown his age, it might have described the place of his birth, or, at all events, his then residence. But, in the absence of this document, we gather from the tombstone in Irthington churchyard some facts connected with his marriage which deserve consideration with reference to his presumed age. In the first place, presuming as we may, from the birth of the eldest son in 1760, that Bowman married in 1759,[1] he was 54 years

---

[1] I am aware Dr. Barnes, writing in 1821, says Bowman married in 1755, when he was fifty years of age ; but if so, it is curious that so many years should have elapsed before the birth of his first child, who, according to one account, was born in 1760, and to another in 1761. The births of the other children followed at short intervals.

of age, while his wife, born in 1726, was 21 years
younger, being only 33. I do not know whether the
yeomen of Cumberland marry young or not, but 54
is, as a general rule, so exceptional an age for a man
to marry at, that the statement is calculated to increase
rather than to remove my scepticism.

'But is not a clue to the absence of all evidence to be
found in a fact which Mr. Gilpin passes over slightly, and
on which his information is probably imperfect. "Bowman," says Mr. Gilpin, "having passed his whole life in
the neighbourhood of his birthplace—*except a few early
years spent in Northumberland.*" Now may not *all* his
early years have been spent in Northumberland (where,
if we knew the precise locality, both his baptismal and
marriage certificates might be discovered), and he have
removed to Irthington on his marriage?

'What was the maiden name of Bowman's wife? where
were their children born and baptized? for the accounts
of Bowman's children are very contradictory. Dr. Barnes,
writing in 1821, says, "he married at the age of 50"
(which would be in 1755) "and had six sons, all of
whom are now living; the eldest is 59 and the youngest
47," which makes the birth of the eldest son to have
taken place in 1761, whereas on the tombstone erected
in Irthington churchyard the eldest son is described as
having "died July 29, 1844, aged 84 years"; according
to which he must have been born in 1760.

'I am writing just now under great disadvantages, and
indeed should not have written at all, but that I feel it
is due to Mr. Gilpin to acknowledge the pains he has

taken to ascertain the truth, but as in my opinion Mr. Gilpin's evidence does not sustain his belief that he has established the fact that Bowman was 118, I feel bound to point out where I think it defective.

'Mr. Gilpin's generosity has, I think, tempted him to take the weaker side ; but whatever may have influenced him, he now deliberately avows his belief that Robert Bowman reached the very exceptional age of 118. I do not say he did not, but I do say there is at present not a particle of evidence that he did so. Those who support the argument that Bowman was 118 must prove their case. "Eo incumbit probatio qui dicit, non qui negat," says the civil law ; and it may be added that the civil law also required that in proportion as the supposed fact was, as in this case, exceptional and beyond the ordinary nature of things, so ought the evidence in support of it to be clear, distinct, and beyond all doubt.'

And so the matter rests. The Rev. Canon Harcourt had unfortunately died (December 10) before the appearance of Mr. Gilpin's report, or of my rejoinder. Mr. Gilpin, disgusted, probably, with the scanty harvest which he had gleaned after all his labour and trouble, left me in possession of the field. My hopes (one of my inducements for writing this letter) that some fresh volunteer might be found to pursue the inquiry, were doomed to disappointment ; and it now remains for those who think that a baptismal certificate of *a* Robert Bowman, baptized in 1705, not proved in the slightest degree to be that of the Robert Bowman, who died in 1823, is evidence of the latter being

118 years, to believe it. I do not: and in the absence of direct and more satisfactory evidence to the contrary shall continue to assert that ROBERT BOWMAN WAS NOT 118.

FREDERICK LAHRBUSH, *not* 106.

Such of the readers of the 'Standard' as are lovers of the marvellous had a dainty dish served up to them on the 24th March, 1870, in a letter from the able correspondent of that journal at New York, containing an account of a public breakfast given in that city to celebrate the 104th birthday of a supposed centenarian. ' But in justice to all parties I will give the account in full.

'Very few men celebrate their birthday in its one-hundred-and-fourth anniversary. But a breakfast-party was given in this city on the 9th inst., to a man who was born in London on the 9th day of March, 1766. When we think of what this man may have seen, of the comparatively remote period of political and literary history with which he forms a connecting link, we are lost in astonishment. All, or nearly all, that Lord Macaulay said of the wonderful lifetime of the poet Rogers, who was born only two years and eight months before our centenarian, and who has been dead more than fourteen years, might be said of him. As a child he may have been patted on the head by Oliver Goldsmith, and may have seen David Garrick act. He may have followed the drums of the regiments, as they marched through London streets, that were sent to reinforce General Gage before Bunker Hill. Many a time he must have passed

Samuel Johnson in Fleet Street. But when we think, not what this man may have seen, but what he actually has seen, our astonishment is even increased. His name is Lahrbush. He entered the British Army on the 17th October, 1789 ; served with the 60th Rifles under the Duke of York in the Low Countries, in 1793 ; was present on the 8th September, 1798, when the French General, Humbert, surrendered to Lord Cornwallis at Ballimanuck, in Ireland ; was with Nelson, in 1801, at the capture of Copenhagen ; witnessed the famous interview between Napoleon and Alexander, which led to the peace of Tilsit, in 1807 ; fought under the Duke of Wellington in the Spanish Peninsula, in 1808–10, displaying such gallantry against Massena at Busaco as to secure a promotion ; was stationed at the Cape of Good Hope in 1811, and distinguished himself in the first Caffre war; and in 1816–17 was an officer of the guard that had the custody of the Emperor Napoleon at St. Helena. *After a service of* 29 *years he sold out his captain's commission in the* 60*th Rifles, in* 1818, and subsequently went to Australia as superintendent of the convict station at Bathurst; in 1837 he removed to Tahiti, from which island he was forcibly expelled by the French in 1842, in consequence of having taken warmly the side of the Protestant missionaries in a controversy with Papal propagandists. For several years he travelled extensively on the Continent. In 1847 he went to take charge of Lord Howard de Walden's estates in Jamaica, but, disgusted with the disorganisation of labour that followed the liberation of the slaves, he came

in the following year, at the age of 82, to this city, where
he has ever since had his abode.   He brought with him
to New York his widowed daughter and grandson, both
of whom soon died, and for nearly twenty years the poor
childless old gentleman has lived quite alone, in the
enjoyment however of wonderful health, in the full
possession of all his faculties, and the vigorous use of
his limbs,

    ' Captain Lahrbush is a good Churchman, and regu-
larly attends morning services on fine Sundays at the
Church of the Ascension in the Fifth Avenue.   An arm
chair is placed for the old captain in the middle aisle,
just in front of the chancel rail, which he occupies
by courtesy of the churchwardens.   He goes through
the forms of kneeling and standing with something of
military precision, and his voice, piped in a high treble,
may be heard in the responses above the rest of the
congregation.'

When I read this I cannot say what astonished me
most, the 'old soldier's' abnormal longevity or his extraordinary adventures, his 104 years, or his presence at the
interview between Alexander and Napoleon.

It was almost as difficult to discredit a statement so
full of precise facts and dates as to believe it.

I therefore looked into the case, and the result of my
inquiries appeared in 'The Standard' of the 2nd April
as follows :—

'Captain Lahrbush is said to have been born on the
9th of March, 1766, in London.   As no parish is speci-

fied I cannot test this, and postpone for the present any observations upon that part of the case.

'Captain Lahrbush is said to have "entered the British army on the 17th of October, 1789." As I gather from the whole tenor of the account that Captain Lahrbush served as an officer, this statement implies that his commission bears that date. I have therefore examined the " War Office Gazette " of that date, but among the 24 officers named in it the name of Lahrbush is not to be found.

'We are next told "he served with the 60th Rifles under the Duke of York, in the Low Countries, in 1793." On this I have to observe that the 60th Regiment was then called the American regiment, and not the Rifles, and that the name of Lahrbush does not appear in the " Army List " for that period.

'Captain Lahrbush, we are then told, "was present on the 8th September, 1798, when the French General Humbert surrendered to Lord Cornwallis, at Ballinamuck, in Ireland : and with Nelson in 1801, at the capture of Copenhagen." Still the name of Lahrbush does not appear in the " Army List."

'I pass over, but wonder how it could have happened, the statement that "he witnessed the famous interview between Napoleon and Alexander, which led to the peace of Tilsit, in 1807," that I may come to the next testable assertion, namely, that " he fought under the Duke of Wellington in the Spanish Peninsula, in 1808-10, displaying such gallantry against Massena at Busaco as to secure a promotion." This promotion after Busaco

suggested to me whether the alleged Centenarian had
risen from the ranks, and gained at this time a commission by his gallantry.   But the "Gazette," which records
the promotions after Busaco, makes no mention of Lahrbush's good fortune, and the "Army List" refuses its
sanction to such theory, for the name of Lahrbush is still
absent from it.

'I have nothing to say about Captain Lahrbush's
services at the Cape of Good Hope, or the manner in
which "he distinguished himself in the first Caffre war;"
or, in 1816-17, "as an officer of the guard that held the
custody of the Emperor Napoleon at St. Helena;" but,
with regard to the next statement, that "after a service
of 29 years he sold out his captain's commission in the
60th Rifles, in 1818," I have this to remark—that with
its persistent disregard to Captain Lahrbush's merits and
services, the "Army List" passes over his retirement
from his regiment with the same mysterious silence
which it has observed with respect to his entrance into
and services in it.

'After this explanation your readers will probably not
be disposed to give credence to the very startling assertion with which the narrative opens, that "a breakfast-
party was given in New York on the 9th instant to a
man who was born in London the 9th day of March,
1766!'

I had no sooner published this letter than I was taught
the wisdom of Lord Melbourne's suggestion, 'Had we
not better leave it alone?'

I had laid myself open to a 'snubbing' and I got it

right and left, from two correspondents, ' Audax ' and
' Anglus,' who, in ' The Standard ' of the 9th of the same
month, thus reproved me for ' riding my favourite hobby
to death ':—

' To the Editor of the Standard.

' Sir,—With reference to a letter which appeared in
your edition dated Saturday, April 2, and signed "W. J.
Thoms," I beg to inform that gentleman, through the
medium of your valuable journal, that on reference to an
" Army List," dated War Office, July, 1815, I find
Captain Lahrbusch serving as lieutenant in the 60th
Royal Americans (now 60th King's Royal Rifle Corps),
date of promotion 29th October, 1809.

' Maxwell's " History of the Irish Rebellion, 1798,"
proves that the 60th Royal Americans served in Ireland
during that eventful period. Captain Lahrbusch might
have then been in the ranks, and subsequently promoted,
which may account for his not being officially recognised
at the time, as it appears nearly impossible, if he entered
the service as an officer, he should after 29 years of service be only captain in those days of hard fighting, when
the 60th most certainly bore their share.

' As to the error of your American correspondent calling the regiment 60th Rifles instead of Royal Americans,
that is very easily accounted for; it is so long since their
title was changed they are more commonly known at
this day by their present distinguished appellation,
though in the interval between their old and the present

designation they were named the Duke of York's Rifle
Corps.

' I beg to remain, Sir, yours, &c.,

' AUDAX.'

' Barnet, Herts, April 5.'

' *To the Editor of the Standard.*

' Sir,—In your issue of Saturday, Mr. W. J. Thoms
comments on the notices which have recently appeared
regarding the alleged centenarian Captain Lahrbush, of
New York. Your correspondent seems to have examined certain old Army Lists, and not finding the old
veteran's name therein, dismisses the whole story, including Captain Lahrbush's very existence, as a myth.

' Is not this something like a riding, on Mr. Thoms'
part, of a favourite hobby to death ? Do the references,
bonâ fide made with every care, exhaust the question ?
Were Army Lists in those days as detailed and correct
as in the present ? Did they contain the provincial
militia corps raised in the several colonies ? Does the
absence of Captain Lahrbush's name establish more than
that there has been an erroneous description of his military service ? Why should he be a Mrs. Harris ? or
why may he not be a very old man because his regimental
services are not properly recorded in a convivial notice ?

' Your readers will draw their own inferences when I
say that, having lived and travelled much with New
Yorkers during the last five or six years, the name of
Captain Lahrbush was repeatedly mentioned as that of
an exceptionally old (centenarian) British officer resident

in New York. He was visited by the Prince of Wales, who gained golden opinions from the New Yorkers by his kindly attention to the old soldier.

'There must be many Americans in London and Paris well acquainted with Captain Lahrbush, to whom I beg to hand over Mr. Thoms ; meantime I enclose my card, and remain your obedient servant,

'ANGLUS.'

How far I succeeded in justifying myself in the following reply, which appeared in the 'Standard' of the 11th April, it will be for the reader to judge.

' *To the Editor of the Standard.*

'Sir,—" Audax" has caught me tripping, and " Anglus" has exactly hit upon the cause of my stumble when he says "Army Lists in those days were not as correct and detailed as in the present." The fact is, finding no trace of a Lahrbush in the 60th Regiment in the earlier Army Lists, I contented myself with referring in the latter volumes to the indexes only, and by so doing have laid myself open to " Audax's " criticism.

'But, having acknowledged this oversight, I hope your correspondent will not think me obstinate if I still adhere to all I have said in my former letter as to Lieutenant Lahrbush (as he was never captain) and his services up to the battle of Busaco, and his alleged consequent promotion ; for since I wrote that letter I have been informed that the name of Lahrbush does not occur in the prize lists for the Peninsula.

'The fact is, the name of the alleged Centenarian first appears in the Army List of 1810 as an ensign in the 60th Regiment — Larbusch (not Lahrbusch), his commission being dated 16th of November, 1809, but neither Lahrbusch nor Larbusch appears in the index, nor, indeed, does De Lahrbusch.

'In the 1811 Army List he appears as a lieutenant, with the name of Lahrbusch, his lieutenant's commission being dated October 29, 1810. Although promotions in the 60th are gazetted in the "Gazette" of the 23rd October, in the "Gazette" containing the first promotions in that regiment after Busaco, no mention is made of Larbusch, neither is there in the "Gazettes" of the 30th of that month or of the 17th of November. His promotion to a lieutenancy, together with that of eight other ensigns of the same regiment, was gazetted the 12th of February, 1811, and was dated War Office the same day.

'Soon after this the name of this officer assumed a fourth form, "De Lahrbusch," and as such it appears both in the Regimental List and in the index of the Army Lists; and I wonder that "Audax" did not see in this change the cause of my error, for the fact is that when "Audax" says that in the 1815 list he finds "Captain Lahrbush" serving as lieutenant in the 60th, he is in error, for the name is "De Lahrbusch," and he is indexed under letter D, as "*De* Lahrbusch."

'I now come to a part of the question which I would gladly have been spared. With a view of testing the statement that "after a service of twenty-nine years he sold

out his captain's commission in the 60th Rifles in 1818,"
I naturally turned to the " Army List " of 1819, and to
the division of it which records the "casualties since the
last publication," and not finding the name of Lahrbush
among the resignations and retirements, I alluded to the
" Army List's persistent disregard of his name and " services." Had I turned the page, however, I should have
read an entry which would have prevented my entering
upon the question at all, and which I only now refer to
because I am not at liberty to suppress the truth ; under
the head of " Cashiered" the reader will find the name
of " Lieutenant De Lahrbusch, 60 F."

' I will make but one further allusion to this painful
part of the subject. In writing to the War Office in
1846 on the subject of his removal from the service,
Lieutenant Lahrbusch, spoke of the cause as his
" youthful errors." They very possibly were so ; and he
may at this time well deserve all that " Anglus " says of
him ; but with reference to the question of his age, I
must remark that if born in 1766 he was fifty-two in
1818 — a time of life which certainly does not justify the
plea of " youthful errors."'

' Surely after this I am justified in repeating, in the
absence of further evidence, my conviction that the
breakfast given in New York was not " given to a man
who was born in London on the 9th day of March,
1766."'

Neither Audax, nor Anglus, nor any other correspondent ventured to impugn my facts or my deductions.
But I received a letter from an anonymous cor-

respondent, commencing, 'In the interest of truth and from a love of fair play, I think it right to tell you, that in . this Lahrbusch controversy you are perfectly right in the main and as far as you go . . . He has added from fifteen to eighteen years to his age, and his birth in London —he is a *born* German ;' the writer adds many curious particulars, and says that if I were hard pressed in the controversy I should hear from him again.

There the matter rested until the 6th May, when the following communication from the New York Correspondent appeared in the same journal :—

' I have seen in the " Standard " of the 2nd April, the letter of Mr. William J. Thoms, calling in question not only the facts connected with the military career of Captain Lahrbush, as set forth in my letter of the 12th March, but the age of that veteran. I can only say in reply, that in New York Captain Lahrbush's wonderful adventures, and yet more wonderful longevity, are universally accepted for truth by the very best people, and that some unkind doubts and suspicions that were expressed several years ago as to his rank and age, were set at rest by the inquiries that were made, with the most satisfactory results, at the proper offices in London. The dates and incidents given by me were all embodied in an article from the pen of General James Watts De Peyster, of this city, which may be found in the " Historical Magazine and American Notes and Queries" for the month of April, 1867. Your correspondent has little doubt that Mr. Thoms may find this work in the reading-room of the British Museum, or at least that it is contained in the library of that institution, and may be consulted upon application to any one of the librarians. If I had the pleasure of knowing Captain Lahrbush it would be an easy matter to ascertain from him the parish in which he was born, but it would be little short of an impertinence in a stranger to call on so aged a man for this information, when the only reason he could assign for so doing was that the Centenarian's veracity had been assailed. Captain Lahrbush is in the full possession of all his faculties, and has been for many years a communicant of the Episcopal Church, and his religious character repels the idea that for empty notoriety he has falsely stated his soldierly experiences and the date of his birth. Englishmen of high social position visiting New York have called on Captain Lahrbush, and his Royal Highness the Prince of Wales when in this city during the autumn of 1860 either entertained the old captain at lunch or met him at lunch at the house of some hospitable citizens, and must have been cognisant of his reputed history. For the rest I may say that the most distinguished men in America are the friends of Captain Lahrbush. Admiral Farragut and General George B. M'Clellan are frequent visitors at his house, and he enjoys in his lonely old age the respect and sympathy of the whole community. He has outlived all his companions and kindred, and passed " beyond the goal of ordinance," but he has not outlived the kindly consideration which brave natures always secure. I have made this reply to the letter of Mr. Thoms, because it seems calculated to shake the con-

fidence of your readers in the absolute verity of this
correspondence. I may be dull or tiresome, and my
deductions may often be idle or unwarranted, but
nothing has ever yet been given by me as fact except
upon what was believed to be competent authority. For
this personal reference, made in justice to myself, the
reader's kind indulgence is invoked.'

No one could doubt the good faith of the writer of
these remarks, or feel other than respect for the kindly
motives which prevented his applying to 'Captain'
Lahrbush for confirmation of his assertions.

I had said my say ; and having shown what the truth
was, was glad to get rid of a disagreeable subject, and
hoped that the old soldier would cease to thrust his
absurd pretensions before the world.

In the spring of 1871 the New York Correspondent of
'The Standard' good-naturedly bantered me on that
unreasonable old gentleman Captain Lahrbush being so
very unreasonable as to insist on living another twelvemonth, and celebrating his own 105th birthday, and in
'The Times' of that year was a report of his giving a
dinner-party to another Centenarian ; and his name
during the same period cropped up in various books
and papers by writers of scientific eminence as an
instance of abnormal Longevity.

In the spring of 1872 the annual exhibition of credulity and imposture was repeated, and the old soldier's
106th birthday, as it was said, was celebrated by a
public breakfast. This brought me so many inquirers
and communications that I felt it right in the interest

of scientific truth, honesty, and common sense to send the following letter to the editor of ' The Standard,' who inserted it in the paper of April 4.

' *To the Editor of the Standard.*

'Sir,—There is nothing so hard to kill as a lie—no one so hard to silence as an impostor.

> ' Destroy his fib or sophistry—in vain—
> The creature's at his dirty work again.'

Old Geeran, the pretended Centenarian of Brighton, died persisting in his imposture ; and the so-called " Captain " Lahrbush seems determined to do the same, if I may judge from the last of many communications which have reached me since I exposed in ' The Standard ' of April 11, 1870, the utter want of truth in his account of himself.

'An American correspondent has forwarded to me the following cutting from the ' New York Tribune ' of March 13, accompanied by some remarks more strong than complimentary, on the credulity of his country-men :—

'" The annual breakfast given by General J. Watts de Peyster to the indestructible veteran Captain Lahrbush, on their concurrent birthday promises to become a permanent institution. The sixth of the series was given last Saturday, on which day the timeproof soldier attained his 107th year. He appeared so fresh and so tough that it seemed scarcely impossible that he would one day see the figures transposed to 701, in

the delicate bouquet before him, which some lady of
the twenty-fifth century should arrange for the second
Methuselah. The dozen guests were, with few exceptions, generals of the highest distinction ; and yet
one of the oldest of them, whom his soldiers used to call
with affectionate familiarity ' Old Joe Hooker,' was a
baby when Captain Lahrbush was a sexagenarian, and
not one had entered the ˙service when he retired under
the burden of his 70 years. The guests all took their
leave, pledging the gay and hearty old gentleman to be
present a year hence, and especially insisting that he
should keep his health and strength to take part in
the centennial of his young friend the American Union."

' Now, as there is scarcely a word of truth in Lahrbush's marvellous account of his age and adventures—
as that account has imposed upon men of high social and
scientific position (it is Lahrbush who is referred to by
Sir Henry Holland in his delightful ' Recollections' as
' a person still living who numbers 106 years ')—it seems
only just, in the interest of truth generally, and of
scientific truth in particular, that as often as his impudent and unfounded claims are brought forward they
should be denounced and exposed. To show up
Geeran's falsehoods, and to allow Lahrbush's to remain
unchallenged, savours too much of overlooking "the
choleric word" in the captain, and treating it as "flat
blasphemy" in the soldier ; whereas really what is bad
in the private is ten times worse in one who has held
his sovereign's commission, and worse still when it is
associated with a show of religion, as in the case of

Lahrbush, of whom it is said he "is a good Christian, and regularly attends morning services on fine Sundays in the Church of the Ascension in the Fifth-avenue. An armchair is placed for the old captain in the middle aisle, just in front of the chancel rail, which he occupies by courtesy of the churchwardens. He goes through the forms of kneeling and standing with something of military precision, and his voice, piped in a high treble, may be heard in the responses above the rest of the congregation."

'Now, let us test by dates and the "Army List" some of the more striking points in the story of "Captain" Lahrbush, who, according to the writer in the "Tribune," "had retired under the burden of his 70 years" "before 'Old Joe Hooker' and the other generals of the highest distinction present at the breakfast had entered the service."

'Mr. Lahrbush says, but does not produce the slightest evidence in support of his statement, that he was born in London on the 9th March, 1766. I have been assured by one who knew him that he is a German, as his name indicates, and that he was not born in London; and I think I shall prove inferentially that he was born most probably about 1786, instead of 1766, 20 years later than he says. He states he entered the British army in October, 1789. He did not enter it till 20 years later, for his ensign's commission in the 60th bears date 10th November, 1809. He has antedated his commission, as he antedated his birth, some 20 years. The fact that he did not join the 60th till 1809 knocks on the head all his absurd stories about serving with the Duke of York in the Low Countries in 1793, with Lord Cornwallis in Ireland in 1798, with Nelson at Copenhagen in 1801, and of his witnessing the interview between Napoleon and Alexander, which led to the peace of Tilsit in 1807.

' Untrue as is the statement which Lahrbush has made as to entering the service, it is not more so than what he has said with reference to his quitting it ; according to which, "after a service of 29 years he sold out his captain's commission in the 60th Rifles in 1818." Now these three lines contain no less than three gross misstatements :—

' 1. Lahrbush served only nine, and not 29 years. Another error of 20 years,

' 2. He never was a captain, and never had a captain's commission to sell.

' 3. He did not sell out, but was cashiered. In the "Army List" of 1819, under the head of " Cashiered," will be found the name of " Lieutenant De Lahrbusch, 60 F."

' And in connection with this unhappy incident Lahrbush has furnished evidence that his statement that he was born in 1766 is not true. Had he been born in 1766 he would have been 52 in 1818, whereas in 1846, writing to the War Office on the subject of his services, he pleads as an excuse for the conduct which led to his removal "youthful errors." "Youthful errors" at 52 !

' And here I bring to a close a task which has been, I suspect, more painful to me than probably it will be to the "gay and hearty old gentleman," who must have

chuckled in his sleeve when he took leave of General J.
Watts De Peyster, and the other generals of the highest
distinction, who had joined in celebrating—shall we say?
—his 87th birthday, under the idea that it was his
107th!'

And with this I close my account of the most bare-
faced case of pretended centenarianism which has ever
come under my notice ; a case made ten times more
offensive by the manner in which a show of religion
appears to be assumed for the sake of gaining credence
to a series of the grossest misstatements.

RICHARD PURSER, *certainly not* 112.

The case of Richard Purser is one of the most unsatisfactory with which I have yet had to deal. It is
unsupported by a single scrap of documentary evidence ;
but distinctly contradicted by the only piece of such
evidence which as yet I have been able to discover. But
of this presently.

Yet in the face of this, and simply on the strength of
the old man's *alleged* recollections, and of the recollections of a clergyman who died as long since as 1837
(five and thirty years ago), as reported by his surviving
daughters, we are called upon to believe that an old man
' who went out to day labour until within the last seven
years' of his life,' that is until he was 105, and whose
photograph taken on his last birthday (three months
before his death) represents a man who, if eighty, is
remarkably stout and hale for that age, had attained the
UNPARALLELED AGE OF 112 YEARS !

Purser and the believers in his abnormal Longevity have not only disregarded the sober advice of the poet—

If you would have your tale thought true,
Keep probability in view—

but thrust probability, common sense and physiology quite out of sight.

Purser's case was first brought under the notice of the readers of 'Notes and Queries' on February 27, 1864, (3rd S. v. 170) by the following communication :—

'A GENUINE CENTENARIAN.—Reading " N. & Q." I find remarks made on " Longevity ;" and as I am personally acquainted with the following most interesting old man, I venture to send you a few particulars of his case ; and should it in any way interest you, and you like to insert it in your magazine, I hope you will do so. I shall be also very happy to present you with his photographic likeness on glass. His name is Richard Purser, born in 1756, on July 14,—so he will be 108 next July. He is residing at Cheltenham, and has 6s. 6d. a-week allowed him : 4s. 6d. from the parish, and 2s. a-week from the 5l. sent annually by the Queen to the clergyman of the place ; he having satisfied her Majesty as to the correctness of the statement, *and discovered the register.* He is a very good old man, attending his church regularly every Sunday, and sacrament once a month ; and was a regular attendant on the weekly lectures up to the last two years, when he was obliged to discontinue some of his habits. He is hale and hearty, and has all his faculties about him ; and is, in every way, a most interesting person. I visit Cheltenham every spring, and see him almost daily for two months, and have a chat with him. Last spring his legs were bent, and his knees touched, with his two feet bowed outwards ; but he managed to get about for his daily strolls with two strong crutches. He has the most charming countenance, and always looks on the bright side of everything.

'WM. EDWARD BELL.'

In the following December Mr. Bell was kind enough to forward from Gorlestone an excellent carte of Purser, taken by Mr. Ellis, of 5, St. Philip's Terrace, Cheltenham, on or about what was said to be Purser's 108th birthday, July 14, 1864. Mr. Bell described the old man as being then 'in his usual health, and able to get about.'

The next I heard of Purser was from the following letter to 'The Times' of January 2, 1867, from a correspondent signing himself 'Gerontophilos' (to which I have already referred, ante p. 4), in which the writer calls him Percy, speaks of him as being in his 110th year, and adds, '*Proofs of his birth and age* were furnished to the minister of his parish in 1860, and were sent to the Queen, from whom he received a gratuity of 5*l.*

A request that 'Gerontophilos' would furnish some evidence in support of his statement, which I addressed to 'The Times,' was not inserted, and so the matter rested until October 1868, when Purser died, and an announcement of his death was communicated to 'The Times' by the Rev. A. Paul, of Cheltenham, of which letter the following is a copy [1] :—

'DEATH OF THE OLDEST MAN IN ENGLAND.

To the Editor of The Times.

' Sir,—The death of one who there is reason to believe to have been the oldest man in England, cannot but interest the readers of " The Times."

' " Richard Purser, died October 12, 1868, aged 112 years." Such was the inscription on the coffin-plate, and in proof of this instance of extraordinary longevity I am told by the Misses Commeline, daughters of the late incumbent of Redmarley d'Abitot, Worcestershire, that Purser was a native of that parish, and that although the register contains no entry of his baptism, they fully believe the old man to have been of the age stated, mentioning two facts in corroboration, viz., first, that Purser was cowman on the farm at Robing's Wood Hill, near Gloucester, when the Rev. James Commeline, born 1762, was curate of the adjoining parish of Hempstead ; and secondly, that he was working in the dockyard at

---

[1] In the ' Introduction' by Dr. Tuthill Massey to the second edition of ' The Life of Thomas Geeran,' that gentleman gives an account of Purser, taken apparently from this communication; but it contains a passage (a very important one, I think) which I presume Dr. Massey derived from some other source. The passage comes after that which says, ' He went out to day labour until within the last seven years, and looked hale and ruddy,' and is as follows: ' His portrait before me exhibits a peaceful happy expression in his face, *looking not more than* 70 *or* 80 *years of age.*' If this statement is Dr. Massey's own, we are, for once, agreed.

Sheerness in the year the 'Royal George' was sunk, 1782. For the last half century Purser has lived in Cheltenham, working as day labourer, and during the last five years her Majesty's bounty has been extended to him in consideration of his extreme age and excellent character by an allowance of 5*l.* per annum. To a friend of mine who questioned him some eight years since, in order to test the reality of his reputed age, the old man said he remembered, when a child of four years old, being taken by his mother to see an illumination in honour of the coronation of King George III. This was in 1760. My friend, who has been his near neighbour for the last 40 years, further tells me that during that long period there was little change in his personal appearance ; indeed he has gone out to day-labour until within the last seven years or so, and looked hale and ruddy to the last. Purser leaves a son, aged 63 years, and his state- ment that he was fully 40 when he married seems to bear out the other statements. He retained his faculties to the last, taking a final leave of his son within an hour of his death. I have reason to know that his wants, temporal and spiritual, were kindly ministered to by the Rev. Mr. Hutchinson, the clergyman of the district (St. Philip's, Leckhampton), and by other benevolent persons, who soothed, by acts of kind attention, the last days of a life drawn out to such a remarkable extent beyond the usual span of human existence.

' I am, sir, yours faithfully,

' ANDREW PAUL.'

' Cheltenham, October 19.'

The persistency with which old Purser's supporters (of whose good faith in the matter I do not entertain the slightest doubt), continued to insist on his having reached such an abnormal age, determined me to look into the matter, but it was not until the close of 1871 that I took it in hand.

I then addressed a letter to the Rev. E. H. Niblett, rector of Redmarley, requesting his good offices in ascertaining, if possible, the real facts of Purser's age. He was from home, on account of indisposition, but wrote me very fully, saying he had received many communications on the subject—that he had searched the registers in vain—detailed some particulars of an interview with Purser's patroness, Miss Commeline, who was very indignant with him for doubting Purser's being so old as he was represented, and referred me for information to the Rev. C. Longfield, his curate, then residing at the rectory at Redmarley. I fear the many inquiries I have since addressed to the latter gentleman must have sorely tried his patience, although they have not exhausted his courtesy.

Mr. Longfield examined the registers from 1753 to 1790, without finding an entry of the baptism of Richard Purser. But there are *lacunæ* in the register. Thus, from 1762 to 1765 there is no entry of a baptism, but there is the mark of a leaf having been removed ; and again there is no record of a baptism for four years from 1785 to 1789.

Mr. Longfield was good enough to make a second

search to see whether by error he had been baptized by
the name of ' Percy' instead of Purser, as he was styled
by 'Gerontophilos,' but the result was the same.

It is clear, therefore, that no evidence of Purser's age
exists at Redmarley ; and the assertion that proof of it
had been laid before Her Majesty is clearly without
foundation. But I am in a position to say more than
this ; for I know, on the best authority, that the clergyman
upon whose application the annuity of 5*l.* was granted
by the Queen to Purser, stated he believed him to be
105, admitted that he had no proof of the fact, and that
he derived his information from Purser himself, and had
no reason to doubt it.

I have been taken to task by the Editor of the 'Wilts
and Gloucester Standard ' on two or three occasions for
doubting Purser's age—in the face of the reported recollections of Mr. Commeline. I have no doubt the ladies
who have handed down these recollections have done so
with the greatest good faith and a perfect conviction of
the accuracy of theirs and their father's recollections ;
but personal recollections *unsupported by collateral evidence are of little worth,* and I am bound to confess I do
not see any such evidence in this case, but have seen
some to contradict it ; and two or three errors of date in
my opponent's charges against me, and two or three
admissions, seem to me to cut the ground from under his
feet. Let me quote one passage from the first of these
articles :—

'We happen to be able to supply some particulars
respecting this old man, who died at Cheltenham a few

months ago, at the reputed age of 111, and whose case
has been referred to in the " Times " between Mr. Thoms
and Dr. Massey. Whether 111 was his right age or not
it is impossible to say, as there is no register of his
baptism, but it must have been so within a year or two,
say *five* years at the outside, still leaving a fair margin
wherewith to establish his claim to centenarianism. The
absence of the baptismal register is no very great loss ;
for all who have been engaged in pedigree-hunting know
how little reliance is to be placed upon a mere entry in
a parish register when unsupported by other evidence.
All that it states is that A. B., son of C. and D., was
baptized on such a day, but it frequently happens that
another A. B. turns up ten or twenty years after the
other, and it is not always possible to tell which is the
one we want. Purser's age, however, is proved in this
way :—A former rector of Redmarley d'Abitot, the parish
where Purser was born, had his title for orders in the
parish of Hempstead, near Gloucester, at which time
Purser was a cowman at the farm on Robin's Wood Hill,
the two men being within a year or two the same age.
Now Mr. Commeline, the gentleman referred to, died in
1837, in his 76th year, having therefore been born in 1762,
and Purser, consequently, must have been born about the
same year, for two young men of 23 could not suppose
one another to be of the same age if one of them was a
boy of 13, as Purser must have been at this time to bring
his age under the 100 ; nor was this merely a casual acquaintance, but it was maintained till Mr. Commeline's
death in 1837, and afterwards continued by the family

till Purser's death, as a Redmarley parishioner. We
have, therefore, here the case of a man whose whole life
from quite a young man was known to the same family;
and although this evidence does not establish Purser's
exact age, it at any rate, renders it quite impossible that
there can have been a difference of 30 years between his
reputed age and his real age, as Mr. Thoms thinks, for
he puts him at 80 at the time of his death; for it is very
clear that he would, in that case, not have been born
at the time that Mr. Commeline was curate of Hempstead ; but there he was, bodily in the flesh, working as
a farm servant, of apparently 22 or 23 years of age. It
is also said that Purser was working in Sheerness dockyard at the time the " Royal George " went down at Spit-
head.　This event happened on August 29, 1782, and
Purser must have been an able-bodied labourer. We
have only to add that the gentleman who has given us
these particulars obtained them direct from à member of
the late Mr. Commeline's family.'

Now on this I would remark that Mr. Commeline died
in 1837, not in his 76th but 74th year, and consequently
was born in 1763—seven years later than Purser claimed.
If then ' the two men' were 'within a year or two the
same age,' what becomes of Purser's recollection of the
illuminations for the coronation of George III. in September 1761 ?　I say nothing of where a Redmarley
baby was to be taken to get a sight of them.

In another article (March 2, 1872), called forth by a
few remarks I had made in ' Notes and Queries ' of
February 10, 1872, the same writer says, ' All we contend

for is that he (Purser) could not be far short of it (112) say four or five years at the outside.'

But take away these four or five years, and what becomes of the old fellow's birthday—July 14, 1756— and his recollections, &c., and the whole superstructure necessarily falls to the ground ; and the writer and myself agree that old Purser was *not* 112, and only differ as to how far he was short of it.

Upon this point the document which I have lately discovered throws no small light.

In the course of my various inquiries into Purser's history I have ascertained two facts. The first is, that he was an illegitimate child, the son of a well-to-do builder named Loveridge, and who had a brother, a solicitor in London ; but whether Loveridge the builder carried on business in Cheltenham, Gloucester or London, all of which have been stated to me, is uncertain. Further information on this point might prove the means of fixing the date of Purser's birth.

The second piece of information which I acquired was, that Purser, who said he ' was fully forty when he was married,' was married at Redmarley.

Again I put the good nature of Mr. Longfield to the test ; and almost by return of post received from that gentleman a copy of the entry of Purser's marriage.

The reader will remember that in the register of baptisms at Redmarley there are two periods during which no records of baptisms are to be found—the first being the period between 1762 and 1765, and the second that between 1785 and 1789.

Now I have already stated my belief, in which I am supported by physiologists and Fellows of the Royal Society,[1] that the photograph of Purser taken about the time of what was said to be his 112th birthday, represents a man only somewhere about 80 years of age, and be it remembered that either Mr. Paul or Dr. Massey, both firm believers in Purser's 112 years, in referring to his death expresses this same opinion: 'The picture before me exhibits a peaceful happy expression in his face, *looking not more than* 70 *or* 80 *years of age.*'

I also believe that it is very rare for men in Purser's condition of life to be so deeply impressed with the doctrines of Malthus, as to put off their marriage till 'fully forty,' yet even supposing that Purser did so—then as he married Ann Rollings on Sept. 12, 1808—it is clear on his own showing he was *born about* 1768 and not in 1756—which *reduces his age by twelve years*; while if, as is much more probable, he was not much more than twenty when he married, it brings him to a little more than eighty at the time of his death, and his birth probably between 1785 and 1789 — the very period during which the baptismal entries are wanting at Redmarley.

Looking to the fact that no person has ever yet been

---

[1] One sets him down as 'between 70 and 80;' a second writer, 'I have not seen any man of 80 who looks so young as the old man whose photograph you enclose. . . . He is very stout and has good stout legs for a very aged man.' A third says, 'He looks so well nourished that I can hardly think him 90. Look what calves and knees he has. Men of 90 generally get to be skinny; and he is plump and in good liking.' I ought to add that in sending the Photograph to my friends, I made no mention of the age of the old man, and did not even state his name.

proved to have reached an age at all approaching to 112, to the physiological evidence against his story furnished by his photograph, and to the unquestionable fact that the only piece of documentary evidence in existence directly contradicts his own story, is it possible to believe otherwise than that Richard Purser was much nearer 80 than 112?

WILLIAM BENNETT, *not* 105, *but* 95.

Nothing could appear more precise and conclusive than the following paragraph from 'The Irish Times' of January 24, 1872, copies of which were kindly sent to me by several correspondents:—

'DEATH OF A VETERAN CENTENARIAN.—William Bennett died at Inchicore, at the house of his son-in-law, James Harrison, on the 23rd inst., at the advanced age of 105 years. He was born in Newmarket, Norfolkshire, in the year 1766, and enlisted in the 32nd Regiment in the year 1793. He was stationed in Ireland previous to the Rebellion of 1798, and served in the Peninsular War under Sir John Moore, John Cathcart, and the Duke of Wellington. He was at the battle of Corunna, and was one of those who assisted at the burial of Sir John Moore. The deceased was discharged from the army, on the pension of one shilling per day, in the year 1814, in receipt of which he continued up to the day of his decease. He retained all his faculties, and enjoyed good health up to the moment of his death.'

Yet that a man should reach the unexampled age of 105; and moreover retain all his faculties, and enjoy good health up to the moment of his death, was so improbable that I could not resist applying for official confirmation or correction of the statement.

The reader will, doubtless, anticipate the result. The records of Chelsea Hospital show that William Bennett, born at Newmarket, Suffolk, enlisted into the Cambridge Militia at the age of 20 years on October 10, 1797, into the 46th Foot on March 18, 1799, and into the 32nd on June 15, 1803.

He was discharged from the 32nd Foot on August 18, 1814, on account of 'general debility;' the period of service which he was allowed to reckon being fourteen years and six months.

When discharged he was about 37 years of age, which would make him about 95 at the time of his decease ; and this is confirmed by the fact that in January, 1862, when applying for an increase of pension, he stated he was 85 years of age.

So it is clear that after all this 'Veteran Centenarian'—as he is styled (whatever that may mean)—was not 105, when he died, but ONLY 95.

MARY HICKS, *not* 104,. *but* 97.

At the close of the year 1870, the following paragraph went the round of the papers :—

'FUNERAL OF A CENTENARIAN.—At the weekly meeting of the Brentford Board of Guardians yesterday, the master of the house reported that Mary Hicks, aged 104, belonging to the parish of Isleworth, died on the

24th November last. She was the widow of John
Hicks, aged 66, who died on the 27th June, 1848. Both
were admitted into the workhouse on the 26th June,
1843. The deceased Mary Hicks was born on the
11th August, 1766, and was baptized on the 15th February, 1767, at Broseley Church, Salop. Since her
admission into the workhouse, now over twenty-seven
years ago, she had fared well, and was a very hale
woman, even after she had lived a century. She retained all her faculties to within a short time of her
death, and would walk about with the aid of a stick.
Her remains were interred in Isleworth churchyard
yesterday afternoon, on which occasion several of the
guardians attended. Four inmates followed, whose
united ages amounted to 335 years (being an average
of 83¾ years), with four other inmates, whose united
ages, added to the above, amounted to 628 years, being
an average for the eight of 78½ years. The scene in
the churchyard drew together a large concourse of
spectators.'—*Daily Telegraph*, Dec. 1, 1870.

I thought the case one which might repay the trouble
of investigation ; and had some intention of undertak-
ing the task. But knowing the amount of trouble it
would entail upon me, and the length of time it would
occupy, I was much pleased to see that the case had
attracted the attention of the 'Daily Telegraph.' In
that paper of December 2 was a short leading article
on the subject of Mary Hicks, which concluded with
the following sensible remarks :—

'In the interest of the integrity of vital statistics we

hold that those who assert that Mary Hicks was a hundred and four years old when she died are bound to prove their statement. First, we want to know whether the registry books at Broseley Church, Salop, state that Mary Hicks was baptized there on the 11th August, 1766; and assuming that such an entry in the register exists, we want to know whether it can be proved that the Mary Hicks so baptized was the same who was admitted to Brentford Workhouse seven and twenty years ago, and died there on the 24th of November, 1870. Unless these things be clearly proved, the story of this "undoubted centenarian" passes into the boundless domain of idle tales. Very old men and women are apt in their dotage to pop an extra decade or so on to their ages, and to this habit most of the stories of abnormal Longevity, which have been demolished by the late Sir George Cornewall Lewis are due. With this we respectfully remit the case of Mary Hicks to the consideration of "Notes and Queries." '

The 'Daily Telegraph' had started the game. 'Notes and Queries' did not follow it up, and it was only lately that my attention was recalled to the case of Mary Hicks, and I renewed my intention of inquiring into it.

I accordingly applied to Mr. Brown, the active and intelligent master of the Brentford Union Workhouse, for information as to the evidence on which the Guardians were satisfied of the great age of Mary Hicks.

Mr. Brown, having put himself in communication with the Guardian who had made all the inquiries, was

enabled, by the courtesy of that gentleman, to forward
to me a copy of her baptismal certificate, of which the
following is a copy. It is an extract from the register
of baptisms at Broseley, Staffordshire :—

'Baptized.—Mary, daughter of Samuel and Mary
Roden, 15th February, 1767.'

As also a certificate of her marriage to her first husband,
John Guest, of which the following is a copy :—

'Married.—John Guest and Mary Roden, May 6th,
1794.'

Mr. Brown also informed me on the authority of the
rector and churchwardens of Broseley that her youngest
sister, Sarah Aston, died in 1861, aged 92 ; and added,
that 'after the death of her first husband she came to
live at Isleworth, and was married to John Hicks ;'
that 'on her admission into Brentford Workhouse, in
1843, she stated her age to be 75, so that she was
entered as born in 1768 — the certificate proved she
was one year older. There was no other Mary Roden
in the family ; and,' added Mr. Brown, 'there can be no
doubt as to the correctness of her age.'

I had some doubts, however, and pointed out to the
Rev. R. A. Cobbold, the Rector of Broseley the grounds
of those doubts, and asked him for some further information—more especially as to whether the Mary Roden,
baptized in 1767, might not have died, and another
daughter, subsequently born, been baptized Mary.'

He very kindly informed me, in reply, that he had
himself carefully searched the register from 1767 to
1780, inclusive, and found no entry of the burial of Mary

Roden, nor of the baptism of Mary Roden other than that of February 15, 1767. (By this, as will be seen hereafter, was obviously meant no other Mary Roden of the same parents.) The first husband of Mary Roden was John Guest, and on May 6, 1794, there is the entry of the marriage—and kindly concluded with an expression of his readiness to afford me any further help in the matter.

I gave up the case in despair, satisfied in my own mind there was a mistake somewhere, but feeling it was not in my power to clear it up. But, to my surprise and satisfaction, this letter was almost immediately followed by another, in which my kind correspondent informed me that the evidence of the age of Mary Hicks had taken a strange turn. That he had on the preceding day seen her nephew, Mr. John Leadbeater, who said that Mary Hicks's father's name was *John*, and not Samuel; that he was not sure of his grandmother's Christian name, but thought it was *Sarah*; that he had given him the names of most of the family ; Mary, the eldest, then two daughters, then Sarah, his mother, who died at Broseley in 1867, aged 93, and then four sons, John, Richard, Thomas, and William. On searching the register of baptisms, Mr. Cobbold found, November 14, 1773, Mary daughter of *John* and *Sarah* Roden.

The register contains entries of the baptism of Anne, on February 4, 1776 ; of Sarah on January 26, 1777 ; of John on February 10, 1779 ; and of Thomas in 1778, among the Dissenters at the end of the year.

Mr. Cobbold comments upon the strange circumstance that Mrs. Hicks should have said that the Christian names of her parents were *Samuel* and *Mary*—and in reply to my suggestion that she did not say so, but it was assumed, when the baptismal certificate of the first Mary Roden, in 1767, was found, confirming her statement that she was more than a hundred,—that gentleman assures me that such was not the case.

The coincidence is certainly curious, but by no means unprecedented ; and furnishes another striking instance of the necessity which I have so strongly urged in the former part of this volume, of ascertaining by preliminary inquiry, both the Christian names and such other particulars of the parents of supposed Centenarians, as will prevent the first baptismal certificate of a child of the name, which is the subject of the search, being at once assumed to be that of the Centenarian. Had it been known that Mary Hicks was the daughter of *John* and *Sarah* Roden, the baptism of the child in 1767 being that of a child of *Samuel* and *Mary* would have been passed by and the search continued till the baptism of the real Mary Roden was found on November 14, 1770, and the age of the old woman settled at once.

While but for the discovery of her nephew (and that Mr. Leadbeater is her nephew there can be no doubt—for he had seen his aunt at Brentford and been recognised by her) and the information respecting her family and identity which he was enabled to furnish, Mary Hicks would have gone to the end of time as having

R

reached the exceptional age of 104 years — more especially as she is necessarily recorded as such on the Registrar General's report, instead of what she really was, ONLY JUST 97.

ADJUTANT PEACOCKE; RICHARD OR WILLIAM TAYLOR; JAMES MCDONALD.

I propose to bring this chapter to a close with three such paragraphs as form the bases of the hundreds of instances of abnormal Longevity recorded in the collections of Easton and Bailey. Nothing apparently can be more clear or convincing than the facts detailed; yet the reader may judge from them what would be the result of applying the tests which I have employed to the long lists of Centenarians detailed in all works of the character to which I have referred.

The first case is that of a veteran officer, communicated to 'Notes and Queries,' of October 15, 1870 (4th S. vi. 317), as follows :—

'Some of your correspondents are interested in the "Centenarian" question. Are any of them aware of the existence of an officer at this present moment who retired from the adjutancy of the 88th Regiment upwards of *eighty-two* years ago, and has been on half-pay since March 31, 1783 ?

'As adjutant he must have served previously for some years to acquire a sufficient knowledge of his duties, and sceptics who doubt whether any one ever really reached 100 years may be convinced by this living in-

stance to the contrary, who must now number considerably more.

B.'

Unfortunately for 'B.,' his 'living instance' has long
ceased to live. It was all very well in the good old
times to appoint mere babies to ensigncies—did not
Mary Lepell, Lady Hervey, get a cornetcy of Dragoons
as soon as she was born?—but, as the correspondent well
remarked, a man must have seen some service before
he got his adjutancy ; and here was a man who had
retired as adjutant from the service for 87 years.
Could he be other than a Centenarian, Yes : Adjutant
George Peacocke, of the 88th Foot, was put on half-pay
April 27, 1783. But when he died nobody knows.
His name not having been struck out of the Army List,
at the time of his death, has been retained there ever
since. He has not drawn pay for so many years that
the attempt to trace when the last payment to him
was made was eventually abandoned as useless.

In the same journal was reprinted from the 'Hull
Observer' of June 20, 1837, the following notice of
Richard Taylor, said to be 104. The heading is
certainly an attractive one.

'DEATH OF THE LAST SOLDIER WHO FOUGHT AT
CULLODEN.—On Friday the 9th instant, Richard Taylor, the oldest pensioner in Chelsea Hospital, was buried
with military honours, in a portion of the ground
attached to the institution appropriated for the interment of old veterans. This mournful but impressive

ceremony drew a vast assemblage of persons present. The deceased was followed by a number of his old companions in arms. He had attained the patriarchal age of 104 years, and his military services comprehended a period of more than fifty years. He was a drummer-boy at the Battle of Culloden in 1745; afterwards he served in Germany under Prince Ferdinand. He afterwards served in various parts of the world. The last action he was present in was on the plains of Alexandria, in Egypt, where the gallant Sir Ralph Abercombie fell. He had been forty years and upwards in the Hospital.'

Anxious to learn the truth of this very interesting statement, I applied to Chelsea Hospital for particulars of *Richard* Taylor. No such name was to be found among the In-Pensioners in June 1837. But there was a *William* Taylor who died in Chelsea Hospital on June 4, 1837; and who, if he was a drummer at Culloden in 1745, must indeed have been a very little one, inasmuch as when discharged on June 17, 1802, Taylor was 62 years of age, and consequently, having been born in 1740, was only five years old when Culloden was fought! He was pensioned from the Independent Companies at 1*s.*; admitted an In-Pensioner December 28, 1806; went out November 1817; again admitted August 1834, and died June 4, 1837, and, which establishes his identity, as I have ascertained from the Register of Burials, now at Somerset House, was buried on June 9, 1837. So Richard Taylor, aged 104, of the ' Hull Observer,' proved to be *William* Taylor of 97!

The third paragraph is yet more extraordinary. The Leeds Mercury of October 19, 1870, contained the following announcement :—

'FUNERAL OF A CENTENARIAN AT PONTEFRACT. —Yesterday there was interred in Pontefract Cemetery a pensioner named James McDonald, who had attained the patriarchal age of 100 years on the 1st instant. Deceased had been blind for nearly four years, and was in receipt of a pension of 2s. per day. He was with Lord Nelson at the Battle of Trafalgar, and was also through the whole of the Peninsular War. He wore three medals, one of which had eleven bars. Several old pensioners followed deceased to his grave. On the coffin was the following inscription :—

James McDonald,
Died 16th October,
1870.
Aged 100 years.

'The Rev. Dr. Bissitt, vicar of Pontefract, read the service.'

My kind friends at Chelsea could give me no information about James McDonald. My mind was made up as to the nature of this statement ; but I made further inquiry. The Staff Officer of Pensioners for the Pontefract district knew nothing of such a man. Then to make assurance doubly sure, and that there might be 'no mistake' as to name, date, or profession, I referred to the Register of Deaths for the period, and find that no death of a person aged 100, at Pontefract, was there registered.

This, then, was not a blunder but a fiction, and yet, from such statements as these I have just quoted, are the popular works on Longevity compiled, without the slightest attempt to ascertain on what foundation the reported cases rest ; and what is worse, once enshrined in these collections, these cases are quoted as authority by grave and scientific writers, when treating on the duration of human life.

CHAPTER IX.

THE cases described in this chapter will serve to show that the principles which I venture to point out as desirable to be followed in investigating cases of alleged abnormal Longevity, are as effective in establishing cases of undoubted Centenarianism as in exposing the untruthfulness of those which have no foundation.

MRS. WILLIAMS, OF BRIDEHEAD, 102.

Mrs. Williams, relict of the late Robert Williams. Esq., of Moor Park, Herts, and Bridehead, Dorset, died at the latter seat on Oct. 8, 1841, at the age of 102 years. She was, according to the inscription on her monument in the parish church, written by her son-in-law, the venerable Vicar of Harrow, the Rev. J. W. Cunningham, 'the youngest daughter of Francis Chassereau, Esq., formerly of Niort, in France (an exile at the age of 14 to this country, in consequence of the Revocation of the Edict of Nantes).'

In communicating to 'Notes and Queries' (2nd S. xi. 58) an account of this lady, her great grandson, Mr. Montague Williams, of Woolland House, Blandford, states that he had 'heard her eldest son, the late Mr. Robert Williams, say that he had dined with his mother on Christmas Day for seventy consecutive years *without a break*—probably an instance *per se* of such a remarkable occurrence in our festive-loving country.'

In January, 1865, an article 'On Longevity' appeared in the 'Quarterly Review' (vol. cxxiv. pp. 179–198) written by a friend of Sir George Cornwall Lewis, who differed very widely in his views upon the subject from those entertained by that accomplished gentleman. Having in 'Notes and Queries' taken exception to some portions of the article in question, Mr. Williams again brought forward, in reply to me and in justification of the reviewer, the case of the lady just referred to. A friendly controversy ensued, in which, I am bound to admit, I came off second best. I do not regret this, inasmuch as my doubts eventually led to the establishment of the truth of a very exceptional and thoroughly authenticated case of Centenarianism.

When the correspondence commenced, neither the place nor the date of baptism of the lady was known, but November 13 was always regarded and kept as her birthday ; and all her family believed her to have been born on that day in 1739, the year she always spoke of as that of her birth. The fourth and youngest daughter of Francis Chassereau, Esq., of Marylebone, formerly of Niort, in France, she was married to Robert Williams, Esq., the well-known banker and M.P. for Dorchester (he died 1814, aged seventy-nine) October 27, 1764, as the entry in her Bible, now in the possession of her grandson, the present Mr. Robert Williams of Bridehead, county Dorset, testifies. Her great grandson has in his possession a large Bible given by her to his father on his twenty-first birthday, in 1820, with his name and an inscription written by her in a very uneven and wandering handwriting ; against which he has put this note, followed by his initials:—

'Written in her 81st year, having the cataract in both eyes.

C. M. W.'

To which he afterwards added below :—

'She was afterwards couched and perfectly restored to sight by Henry Alexander, Esq., on the 22nd of Nov., 1820, being 81 years of age.'

On the opposite page, and two years after, she has again written his name, &c., but now in a good clear hand, having then the use of her sight, which she preserved to the last ; to which he has again added this note :—

'Oct. 1822.   Written in her 83rd year.'

There is also in existence another Bible given by the old lady to the late Admiral Sir H. L. Baker, Bart., in 1830, with his name and the date written by her ; to which she has appended her signature ; under which is written by Lady Baker (Mrs. Williams's granddaughter) this note :—'Written in her ninety-first year.'

As instances not only of her physical powers, but of her vigorous intellect at an advanced age, it may be added, 'that in 1829, being then in her 90th year, she held her great-granddaughter and godchild in her arms at the font,' and that on the occasion of her grandson, the present Mr. Robert Williams, of Bridehead, coming of age, on January 23, 1832, when she was in her 93rd year, when the assembled tenantry and others offered her their congratulations and drank her health, she stood up and herself returned thanks in a not very short speech.

On October 8, 1841, this venerable old lady, for she was not less remarkable for her age and vigour than eminent for the child-like simplicity of her earnest piety, sank to her rest, and on the 15th was followed to the grave by her eldest and only surviving son, then in his 75th year, her two sons-in-law, the late Sir Colman Rashleigh, Bart, and the Rev. J. W. Cunningham, late Vicar of Harrow, and by numerous grandchildren, great grandchildren, and other relatives and friends.

A short time afterwards all possible doubt as to the precise age of this lady, viz., that she was within a month of 102, was removed, by the discovery in the admirably kept register of St. Martin-in-the-Fields (in which parish her father at the time of her birth resided, carrying on business in Long Acre), of the following entry :—

' 1739, Nov. 14.   Jane d. of Francis and Ann Chatte-reau—born Nov. 13.'

Though the name is misspelt, Chattereau instead of Chassereau, there can be no doubt that the entry applies to the lady in question, and the addition of the date of birth (November 13—the day on which she always celebrated her birthday) is a striking confirmation of what had always been said respecting her age.

Although it was improbable that her parents should have two children baptized by the same name and born *on the same day of the month,* viz., Nov. 13, I myself examined the register down to March, 1744, and though I found two entries of baptisms of sons of Peter and Mary Chassereau, there is no subsequent baptism recorded of a child of Francis and Ann.

We are enabled by the courtesy of Mr. Montague Williams to give the following particulars of the life of this interesting lady :—

'She appears to have made codicils to her will in her own handwriting in 1834, and again in 1838—when she was respectively in her 95th and 99th year ; and on the 12th day of November, also in the latter year, when she was within a day of being in her 100th year, she made an alteration in her will of considerable extent, and which was duly acted upon after her death. An old friend of the family writes me word " I do not recollect her memory for events long past (such as visiting in a house on *old London Bridge*) failed until after she had completed her 100th year. At 95 she used to make breakfast for a large party of children and grandchildren, remembering the different tastes of each, from the eldest to the youngest. Her recollection of what she had learnt in her youth, the psalms, ' Te Deum,' 'Magnificat,' 'Nunc Dimittis,' Bishop Ken's Morning and Evening Hymns *in full*, and pre-eminently the Catechism, remained fresh in her memory more or less to the last. Only four days before her death, during a drive of seven or eight miles, she repeated the latter to me.

With those exceptions, her mental faculties had much weakened for two years before; also her powers of walking had failed before that. One thing I must tell you, that, unlike R. Tichbourne, she had not forgotten her French, and she would speak in it when she wished that all present should not understand her, but not during her last year."

'To this I may add that when past 90 she cut a third tooth, which was always a source of inconvenience and annoyance to her; and I well remember her giving me and my sister and other great grandchildren a sovereign each on Christmas day, 1840, being then in her 102nd year, adding as she gave it, " You will not very likely again have a sovereign given you by an old lady of 101." I think these particulars being authentic may interest you, and you are quite welcome to make any use of them you like in your forthcoming work. I should add that the old friend who wrote me the first particulars has often told me that the old lady latterly appeared quite unlike an ordinary being, her flesh and skin appearing so different from that of ordinary old persons.'

MR. WILLIAM PLANK, 100.

The case of Mr. William Plank, who died at Harrow on November 19, 1867, having just completed his 100th year, is as clearly established a case of Centenarianism as can well be; seeing that his age at various periods of his long life is authenticated by official records.

It was first brought under general notice by the following letter, which appeared in 'The Standard' of November 9 of that year:—

'I have thought it worthy of public record that Mr. William Plank, an old inhabitant of this town, has this day attained the remarkable age of 100 years, having still the use of all his faculties, with the exception of that of vision, which he lost eleven years ago. He has been an inhabitant of Harrow, occupying the same house, 56 years. He is the son of James and Hannah Plank, of Wandsworth, Surrey, where he was born on Saturday, November 7, 1767, and baptized November 17 of the same year. It may be of further interest to record that for a year (viz. in 1780) he was a schoolfellow of the late Lord Lyndhurst. They were at the school of Mr. W. Franks, of Clapham. Mr. Plank left in 1781, leaving young Copley still at the school.

'Mr. Plank was originally intended for commercial pursuits, and was bound apprentice at Salters' Hall, City, on March 22, 1782, to his elder brother, a calico printer and a member of the Salters' Company. Mr. Plank is and has been for many years "father" of the Salters' Company. He was admitted to the freedom and livery of the company and the city on October 20, 1789, and therefore may be considered almost to a certainty the father of the city of London. I saw him out walking, with the assistance of a friend, the day before

yesterday, and at his house to-day. He is quite
cheerful, and well able to receive the congratulations
of his friends and neighbours.

' I am, sir,

' Your obedient servant,

' WM. WINKLEY, F.S.A.

' Harrow, Nov. 7.

' P.S.—Before he came to Harrow he was frequently
ailing.'

On reading this letter, and seeing how satisfactorily
the case might be established, I first instituted a search ·
in the Baptismal Register of Wandsworth, where his
baptism is registered as follows : ' William, son of James
and Hannah Plank, christened November 20, 1767.' · I
then applied to Mr. Overall, the active and intelligent
librarian of the City of London Library, who kindly
procured for me the following information, viz. :—

That William Plank was apprenticed to Mr. James
Plank to learn the trade of a calico printer on May 28,
1782, at which time he must have been upwards of
fourteen years of age.

That his indentures bear the following endorsement :
' Took up his freedom in the Salters' Company, October
20, 1789.' At which time, as no one can take up his
freedom until he is of age, Mr. Plank must have been
twenty-one and upwards.

Mr. Plank had been for many years father of the
Salters' Company, and at the dinner held by the company

after the monthly court held by them for the transaction of business on November 7, the centenary of Mr. Plank's birth, the company received from him the following telegram :—

'Mr. Plank, Harrow, to the Master Warden and Court of Assistants.

'Mr. Plank has this day completed his 100th year, and in good health and spirits.   A party of friends dine with him to-day.'

To this an answer was returned announcing :—

'That the Company were then drinking the health of their Centenarian Colleague.'

The 19th of November closed the life of the father of the Salters' Company, who survived his Century only twelve days. That he was those twelve days more than a hundred years old may fairly be concluded, as although there is no evidence of the precise date of his birth, there can be little doubt that he was born on November 7, the day which he always kept as his birthday, and which was only thirteen days previous to that on which he was baptized.

MR. JACOB WILLIAM LUNING, 103.

It was while this venerable gentleman was living, viz., April, 1868, that there appeared in 'Notes and Queries' the following interesting particulars of him from the pen of Mr. W. H. Cottell :—

'There is now living at Morden College, Blackheath, Mr. Jacob William Luning, born at Hamelvörden, in the kingdom of Hanover, on May 19, 1767. To enable him to succeed to some property which belonged to his mother, he obtained, *forty one-years ago*, a certificate of his baptism. A verbatim copy is subjoined. Mr. Luning was the elder of two sons ; his brother Conrad died in London nearly fifty years since. He married at Spalding, Lincolnshire, August 4, 1796, Eleanor, daughter of a Captain Sands, and by her had fifteen children. Excepting deafness, Mr. Luning is at this time in full possession of all his faculties of mind and body ; his teeth and hair are comparatively sound and complete ; the latter has, however, been whitened by the snows of one hundred winters. He takes a daily walk in fine weather, and reads without glasses. These aids he discarded on receiving his second sight some ten years since. This gentleman claims descent, through his mother, from Christina, sister to Martin Luther ; and I hope in a short time to be allowed to inspect some family papers said to prove such to be the fact. Should they confirm Mr. Luning's claim, probably a space may be found for his pedigree in "Notes and Queries" :—

' Certificate of Baptism extracted from the Church Books at Hamelvörden therein written in the following words :—

' In wedlock born 1767 (one thousand seven hundred and sixty-seven) the 19th of May, the son of the here resident Clergyman, Meinhard Conrad Luning, and his wife Magdalena Dorothea, born Pratje, baptized the 21st instant, and named Jacob William.

'Witness the Inspector of Customs Mr Luning of Verden—

'That the above is truly extracted I hereby certify by my own handwriting, signature, and seal of office, in fidem—

'FREDERICK DAVID WERBE,

'Superintendent & Clergyman at Hamelvörden in the district of Kehdingen, kingdom of Hanover, the 30th March, 1872.

'(L.S.)'

On May 20, 1869, there appeared in 'The Times' a letter from the Hon. and Rev. John Harbord, of Morden College (called forth by my exposure of the case of an old soldier, who claimed to be 105, but proved to be 80, stating that on the day it was written (19th) a member of Morden College had completed his 102nd year, 'in perfect health, and in possession of all his faculties, though certainly deaf.' Mr. Harbord having offered additional particulars in corroboration of Mr. Luning's age, I invited him to produce them, more especially such evidence as went to prove the identity of the child baptized with the Centenarian of Morden College.

The following is from that gentleman's reply in 'The Times' of May 27 :—

'"Our centenarian," besides the certificate already alluded to, which gives the date of birth as well as baptism, is in possession of a printed book, written by his uncle, and published by him in the year 1784, entitled, "A Short Account of the Life, the Writings, and the

S

Family of the late Superintendent-General P., with his Genealogy;' and in this book occurs the following passage concerning "our centenarian:" "Jacob William first saw the light of this world at Hamelvörden in the year 1767, on the 19th of May. He is intended to be a merchant, and having been duly instructed in Christianity, arithmetic, writing, and other useful things, he has been sent to Hanover, where, under the guidance of the worthy M. von U., he intends to qualify himself for a merchant."

'You will observe that the date of his birth given in this paragraph exactly corresponds with that mentioned in the certificate of baptism which was obtained from the church book at Hamelvörden in 1827.

'In the year 1790 "our centenarian" arrived in London, and became at once a clerk and bookkeeper in a well-known and highly respectable house in the city. It surely is very improbable that this firm should appoint to a responsible post a man with whose antecedents they were not well acquainted, and of whose identity they were not fully assured.

'In 1859 "our centenarian" applied to be admitted into this college, and he was elected a member on the 30th of March in that year, having in his memorial to the trustees, which was attested and corroborated by several respectable merchants, stated his age to have been 91 years on his previous birthday. This again agrees with the age mentioned in the book and certificate of baptism.

'He has, moreover, children, and there are also other

persons still alive, one himself a member of this college and a fellow countryman and brother merchant of " our centenarian," who know him to be a native of Hamel-vörden and the person he represents himself to be. Now, this may not be proof conclusive enough for a court of law, but it surely is strong presumptive evidence, and sufficiently so to warrant the use of the words in my former letter, " undoubted proof," which Mr. Thoms appears to carp at and dispute.'

All this was strongly in favour of Mr. Luning being of the age claimed ; but it was not such evidence as so exceptional a case demanded, as I ventured to point out in my reply to Mr. Harbord :—

' The age of 102 claimed for " our centenarian " is so extremely exceptional that it obviously can only be admitted upon the most conclusive evidence. I am quite willing to allow that the evidence which Mr. Harbord has furnished to your columns to-day is " strong presumptive evidence " in favour of his view, but then Mr. Harbord admits that " it may not be conclusive enough for a court of law," and, therefore, I contend it is not such complete and conclusive evidence as such an exceptional case demands.

' There is no question that the gentleman referred to is of very great age, inasmuch as he is said to have been married at Spalding so long since as the 4th of August, 1796, but *non constat* that he is 102 ; and no evidence of his children or of any other one now living can prove his identity with the child baptized in 1767.

'Marriage certificates sometimes record the ages of the parties. Perhaps Mr. Harbord may be able to ascertain whether the certificate of this marriage does so, as such information would have an important bearing upon the case. Mr. Harbord may, perhaps, be able to ascertain from the curious book to which he has referred, the date of birth of the elder brother Conrad, who died in London some fifty years since; and also, if these gentlemen had any sister or sisters, when and where they were born.

'The question is one of great interest, especially in connection with medical science and life-assurance, and I venture to hope, in spite of the many claims upon its columns, " The Times " will find space for its thorough investigation.'

I ought, perhaps, to add in justification of my doubts, that I was in possession of some information better calculated to strengthen than to remove them.

Here the correspondence ended. On June 23, 1870, the long life of this gentleman—103 years one month and four days—came to an end ; and then came out a piece of evidence of the most conclusive kind, namely, that at the age of 36 he had insured his life in the Equitable. No man ever makes himself older than he is when effecting an insurance, and few live seventy-seven years after it.

This remarkable case was invested by the Registrar General, who communicated the following interesting particulars of it to 'The Times' of July 8, 1870:—
'Jacob William Luning, who died on June 23, aged 103

years, at Morden College, Blackheath, was born at
Hamelvörden, in Hanover, on May 19, 1767. He came
to London at the age of 23, and was a boarder at Mr.
Duff's school in Tooting ; he was naturalised, and married
Ellen Sands, at Spalding, in Lincolnshire, in 1796 (age
29) ; insured his life for 200*l.* in the Equitable Society
at the age of 36 ; had twelve children born and christened,
of whom six survive—three sons and three daughters—
of ages ranging from 53 to 66. These children were
born, therefore, when he was between the ages of 37 and
50, from eight to twenty-one years after his marriage.
Not succeeding in business himself, he became bookkeeper in some of the first mercantile houses in the city,
and was engaged in this vocation until he attained, in
1858, the age of 91. He was admitted a member of
Morden College on March 30, 1859, having in his
memorial to the trustees stated his age to have been
91 on his previous birthday. These particulars have
been supplied to the Registrar General by Robert Finch,
M.D., Medical Officer of Health for Charlton, who has
also answered some inquiries and supplied documentary
evidence which satisfactorily establish the facts. Dr.
Finch states, on the daughter's authority, that from the
date of admission into the College until the last few
months the old man enjoyed good health, and, with
the exception of some deafness, was in the possession of all his faculties. His strength gradually gave
way, and for about a fortnight he was unable to leave
his bed. The light that had burnt for 103 years went
out. The father, Meinhard Conrad Luning, pastor of

Hamelvörden, was born on December 17, 1732, and married, on May 25, 1764, Magdalena Dorothea Pratje, born at Stade on January 23, 1748. Jacob William's father was 31, his mother 16, at their marriage ; 34 and 19 at his birth. The father died of bilious fever, aged 51 ; the mother attained the age of 78¾ years. They had four sons, two daughters ; a son and daughter died in infancy, one died aged 22, two grew up. Jacob William was the third child. He is represented in his pedigree as the eleventh in descent from Christina Luther, the sister of Dr. Martin Luther, who died without issue. Dr. Finch cites as his authority the life of Superintendent-General Pratje, the grandfather of Luning. The following document is important :—The verbal translation of the Certificate of Baptism :—Certificate of Baptism, extracted from the church book at Hamelvörden, inscribed in the following words :—" In wedlock, born 1767, the 19th of May, the son of the here resident clergyman, Meinhard Conrad Luning, and his wife, Magdalena Dorothea (born Pratje), baptized the 21st inst., and named Jacob William. Witness, M. Luning, of Verden, Inspector of Customs." " That the above is truly extracted I hereby certify by my own handwriting, signature, and seal of office. In fidem, Fredk. David Werbe, Superintendent and Clergyman at Hamelvörden, in the district of Kehdingen, Kingdom of Hanover. Hamelvörden, March 30, 1827." (L.S.) The referees say the life was good ; he had had smallpox. The bonuses had raised the policy to 1,292*l.* 10*s.* This information is supplied by the Equitable Society.'

One word more. This venerable gentleman, who, it will be seen, attained the exceptional age of 103 years, one month, and nine days, furnishes the only instance out of the thousands, might I not say, hundreds of thousands of assured lives, of any such life being extended to 100 years.

MRS. CATHERINE DUNCOMBE SHAFTO, 101.

MRS. CATHERINE DUNCOMBE SHAFTO, who died at Whitworth Park on March 19, 1872, at the advanced age of 101 years, one month, and nine days, was a very remarkable woman. She *did not remember* walking to church to be baptized. Seeing that she was born on February 10, 1771, and was baptized on the following day the reader may see nothing very extraordinary in this. But if he had had as much to do with ascertaining the real ages of ladies claiming to be Centenarians as I have had, he would have found that in almost every instance the old lady, in addition to the hundred and odd years which she claimed, added a 'bittock' of a few years more, in the stereotyped formula that 'she perfectly recollected walking to church to be baptized.' Mrs. Shafto was the exception which proved the rule; and this perhaps was owing also to the exceptional fact that she really *was* a hundred and upwards.

Only a few weeks before this venerable lady died, a friend placed in my hands, a certificate of which the following is a copy.

'" Parish of Saint Andrew, Auckland. — Baptismal

Register.—February 11, 1771.—Catherine, daughter of Sir John and Lady Dorothy Eden of Windleston."

'Feb. 1, 1872.

' I, Henry A. Mitton, vicar of St. Andrew's, Auckland, do hereby certify that this is a correct extract from the register books of the said parish. Witness my hand this first day of February, 1872

'HENRY A MITTON.

'The above-mentioned Catherine Eden, baptized on February 11, 1771, was married, in 1803, to the late Robert Eden Duncombe Shafto, Esq., of Whitworth, co. Durham, and is now, February 3, 1872, living at Whitworth.

' HENRY A. MITTON,

' Vicar of St. Andrew's,

' Auckland.'

' Feb. 3, 1872.'

The identity between the subject of the baptismal certificate and the Centenarian is in this case much more clearly established than usual. But all doubts upon that point must be effectually silenced by the fact, already referred to in the letter printed (at p. 26) from Sir Alexander Spearman, which records that in October 1790, she, being then in the 19th year of her age, was selected as one of the Government nominees in the tontine of that year.

The deceased lady was the third daughter of Sir John

Eden, Bart., of Windleston, by his second wife Dorothy, only child of Peter Johnson, Esq., of York.

. She married, in 1803, Robert Eden Duncombe Shafto, Esq., who represented the city of Durham in Parliament from 1804 to 1809, and died in 1848, aged 72. By him she had five sons and one daughter, three of the sons, namely Robert Duncombe-Shafto, who represented the northern district of Durham in several successive parliaments and retired at the last general election ; Thomas Duncombe-Shafto, Esq., and the Rev. Arthur Duncombe-Shafto, rector of Brancepeth and rural dean.

Like Mrs. Williams, this lady inherited a strong constitution and a vigorous intellect. Her hospitality was unbounded. She took a lively interest in the welfare of all around, and was ever ready to render assistance in every work of faith, love, and charity. A striking proof of her activity and intelligence at a very advanced age is shown in the fact that on the day on which she completed her 100th year, she appeared both at breakfast and dinner at the wedding of a granddaughter, which took place on that day. During the whole of her long life she had enjoyed the best of health, and retained her intellects unimpaired to the last. Even on the morning of her death she conversed freely with her medical attendant, Dr. O'Hanlon of Spennymoor, and spoke of her death as rapidly approaching.

CHAPTER X.

No reasonable doubt can exist that the four venerable individuals whose cases have been described in the preceding chapter, had really attained the great age claimed for them.

In the cases to which I am about to call attention, the parties may possibly, and some probably, have been as old as they are supposed to have been ; but the evidence in support of their claims is not so clear and continuous. There is no such confirmation of the age shown by baptismal certificate as that afforded by Mrs. Williams's own statement twenty years preceding her death,—by Mr. Plank's apprenticeship, and admission to the freedom of his company,—by Mrs. Shafto's nomination in the Tontine, or by Mr. Luning's insurance of his life when he was 36 years old—absolutely nothing, in short, proving the identity of the Centenarian with the child named in the baptismal certificate—which may be said to be in the following instances the only evidence of age.

Mrs. Lawrence.

The author of the article on ' Longevity and Centenarianism,' in ' The Quarterly Review,' to which I have already referred, was good enough to forward to

'Notes and Queries,' of March 1, 1868, the following interesting case :—

. 'By the courtesy of Major-General Lawrence, of Sydney Place, Bath, I am enabled to offer you a well-attested case of Centenarianism. General Lawrence's mother, Mrs. Martha Lawrence, daughter of John Cripps, Esq., of Upton House, Tetbury, was born on August 9, 1758, in Bow Lane, Cheapside, and christened at St. Mary's, Aldermary.   She died on the morning of February 17, 1862, and was buried in the grave-yard at Ham Common, Surrey, in a grave beyond the church, to the east.   On the tombstone are inscribed the dates of her birth and her death.   Thus she must have attained the great age of *one hundred and three years, six months, and seven days, when she died without a struggle, in full possession of her faculties.*

'General Lawrence informs me that, on a fly-leaf of an old family Bible in his possession, is the following entry :—

"John Lawrence and Martha Cripps were married on the 12th Nov^r, 1783, at Streatham."'

This case is so exceptional as to call for undoubted proof of the identity of the child born on August 15, 1758, with the aged lady who died February 17, 1862.

The only entries in the baptismal register of St. Mary, Aldermary, are :—

1758. Aug. 15. Martha, daughter of John and Frances Cripps.

1762. July 29. Frances, daughter of do.

1764. Jan. 1. Margaret Anne, daughter of do.

I am bound to add there is no evidence of the burial of the child Martha, baptized in 1758.

On the possibility of the register of the marriage containing evidence of Martha Cripps's age at that time, I ventured to trouble the Rev. J. R. Nichols, the Rector of Streatham, with an inquiry. In reply to which he obligingly sent me a copy of the register. In this her age is not mentioned, but she is described as being of the Parish of Clapham.

I am indebted to the courtesy of her son, General Lawrence, for the information that she had four sisters, younger than herself, viz.: Frances, Margaret, Mary, and Sophia. Frances and Margaret were baptized at St. Mary's, Aldermary. It is probable that Mary and Sophia were born after their parents' removal to Clapham. When I add that the entry in the Family Bible is repeated in a book called 'The Mystery of the Soul,' there is, I believe, all the evidence that exists that Mrs. Lawrence attained the exceptional age of 103 years, 6 months, and 7 days, in full possession of her faculties. A lady who died nearly 80 years after her marriage must, at all events, have been nearly a Centenarian.

SALLY CLARK.

The case of Sally Clark, who was buried at Hawarden on April 21, 1871, was brought forward some years previously by Mr. Thomas Hughes, of Chester, in 'Notes and Queries,' of January 25, 1868 (4ᵗʰ S. i. 71);

not, as is too frequently the case, hastily and without full inquiry, but after a patient investigation, which .must have cost him a considerable amount of time and trouble.

'There is now (1868),' says Mr. Hughes, 'living at Hawarden, in the county of Flint, an old lady named Sally Clark, who claims to have been born at Caerwys, in that county, in the year 1762. She reckons her age (106) from the date of her marriage in 1790, at which time, she declares, she was 28 years old. She further declares that she *walked* with her parents to Caerwys Church on the day of her christening. I give these pre-liminaries on the testimony of the good old dame her-self, although it will be seen as we proceed that they require a certain amount of qualification. The actual facts, as ascertained by registers and other documents in my possession, are as follows :—

' John Davies and Rose Roberts were married in the neighbourhood of Mold, Flintshire, and had a first-born daughter, Margaret, living when they migrated to Caerwys in 1757. Other children were born to them there, viz., Elizabeth, baptized in 1757; John, in 1758; Mary, in 1761 ; and Jane, in 1764. And now comes in chronolo-gical order the following document, duly stamped and attested, under the hand of the Rev. W. Hughes, the present Rector of Caerwys :—

' " Baptism solemnised in the parish of Caerwys, in the county of Flint, in the year 1767.

' " Sarah, daughter of John Davies and Rose his wife, baptized the 1st of March.

' " The above is a correct extract from the Register Book of Baptisms belonging to the Parish Church of Caerwys aforesaid.

' " W. HUGHES, Rector of Caerwys.

"January 2, 1867."

' I may add that the baptisms of another daughter, Anne, and of a second son, Jonathan, appear respectively under the years 1769 and 1772.

' When about twelve years old, Sarah Davies left her parents at Caerwys to live as servant on the farm of Mr. Gibbons, of Ewloe town, in the parish of Hawarden. She continued as a servant in the neighbourhood until 1790, in which year, upon March 3, being at the time described as " Sarah Davies, spinster," she was married, " after banns " at Hawarden Church, to " William Clark, bachelor and labourer," as appears by a stamped copy of Marriage Register, No. 319, kindly supplied to me by the Rev. Henry Glynne, Rector of Hawarden. Sally Clark continued to live in the parish of Hawarden until the death of her husband, on January 20, 1844 ; prior to which time she had become the mother of ten children, the youngest of whom is now 57 years of age ; the oldest, a daughter, Mrs. Elizabeth Blundell, aged 77, is now resident with her own family of grandchildren at West Derby, near Liverpool. Another daughter and a son live each in separate cottages on the outskirts of Hawarden ; and along with the last-named, happy and whole in mind, but not of course very active in body, resides our Centenarian friend Sally ; and it is, as I learn from

eyewitnesses, not uncommon even now to see the ancient
dame, who is grown almost blind, sitting in her armchair,
. with one of her many great-grandchildren seated on her
knee.   A short time ago, at the suggestion of Mrs. Gladstone, who is much interested in the old lady, I had a
photograph taken of the worthy matron, sitting at her
cottage door, on the lintel of which, above her head, is
nailed an old horse-shoe, the universal "harbinger of
good luck" all over the world.   Sally Clark has had ten
children, thirty grandchildren, and at least thirty-two
great-grandchildren, most of whom are still living, and
naturally proud of their ancient patriarch.

'It will now appear that, supposing the old lady to
have been baptized on the very day of her birth (which
is not likely), she will be 101 years old if she lives until
March 1 in this present year (1868).   Further than this,
if her statement be correct that she walked to Caerwys
Church to be christened, she would be at least two years
older still!   Her brother John's son, Thomas Davies, is
now, or was very recently, living in the Mold, aged upwards of 80!   Her mother, Rose Davies, and her two
brothers, John and Jonathan Davies, lie buried in the
churchyard at Mold.   Her sister Jane married in Chester, and went to reside at Backford, near this city, where
she died several years ago; and Anne, another sister,
died and was buried near London.'

'Such are Sally Clark's claims to Centenarianism—
claims which I believe both Mr. Gladstone and the
Bishop of Chester, who have examined them, regard as
clearly established.'

When this statement was published, it was suggested to Mr. Hughes that further search [1] should be made, to ascertain whether the Sally Davies baptized in 1767 might not have died early, and another child have been baptized Sarah—a very common case, and one which often leads to unfounded claims to Centenarianism.

It is the more called for in this case, because, if correct in two of her statements, Sally Clark was clearly not the Sarah Davies baptized in 1767. That child, having had a sister baptized in 1764, would doubtless have been baptized at that time had she been born: but if, as is most probable, she was not born for at least a twelvemonth after that time, she could not *possibly remember walking to Caerwys Church* to be baptized: while the Sarah Davies of 1767 was 23, and not 28 (on which point the late Rector of Hawarden informed Mr. Hughes 'she was very positive') at the time of her marriage.

Looking to the very exceptional age claimed for old Sally Clarke, I leave it to my readers to decide how far such age is established by the evidence produced in support of it.

PEGGY LONGMIRE.

My correspondence on the subject of this Westmoreland Centenarian would fill a moderate-sized volume ; and my note on it will, I fear, occupy more space here than I can well afford ; but it is bare justice to the Rev. Mr. Bright, of Windermere, to Mr. W. Jackson Browne, of

---

[1] This investigation, which has only recently commenced, and is not yet concluded, has already brought to light an elder sister, 'Sarah,' baptized in 1762, and a brother, 'Edward,' not previously known.

Troutbeck, and Mr. Somerwell, jun., of Kendal, who have borne with exemplary patience the numerous inquiries with which I have troubled them, that Peggy's case should be somewhat fully treated.

The first notice of her appeared in the spring of 1868, in the shape of the following paragraph, which was then going the rounds of the press :—

'Peggy Longmire, well known in the Westmoreland district, died at Troutbeck on Sunday last, having completed her 104th year on the 15th of April. She was quite a notable character in the district, and was the grandmother of " Tom," the celebrated champion wrestler. Two of her three children survive her, as do also 10 grandchildren, 30 great-grandchildren, and three great-great-grandchildren. Peggy lived for many years near the public-house called the " Mortal Man," with whose quaintly inscribed signboard visitors to the beautiful valley of Troutbeck are no doubt familiar :—

> " O, mortal man that liv'st on bread,
> How comes thy nose to be so red ? "
> " Thou silly ass, that looks so pale,
> It is with drinking Birkett's ale."

Peggy enjoyed her usual robust health until about a month past, during which she suffered considerably, but remained perfectly conscious to the hour of her death. The following character, given to her when she was a servant with Mr. G. Browne, of Troutbeck, shows that she was made of sterling stuff :—" To all whom it may concern. These are to certify that Margaret, the daughter of John Atkinson, of Applethwaite, in the parish of

Windermere, in the county of Westmoreland, served me as a diligent, faithful, and honest servant for two years, viz., from Whitsuntide 1783 to 1785, and that during all the said time I never saw, heard, or had reason to believe but that she was virtuous and modest. Witness my hand this 23rd October, 1788.—GEORGE BROWNE, Troutbeck." '

Shortly afterwards a gentleman sent me a transcript of the notice which appeared in the 'Kendal Mercury' of Saturday, May 30, 1768. I reprint this entire, because, strange as it may seem, after so short a time, some of my correspondents in Westmoreland have been unable to get, or even to consult, the paper in question :—

'She was the daughter of *John and Mary Atkinson*, of Far Orrest, in Applethwaite, in the parish of Windermere, Westmoreland, who were *members of the Society of Friends*; but, occupying a small farm under the then Rector of Windermere, they were induced to conform to the Established Church. This occurred when Margaret was in her thirteenth year (according to the register of her baptism, kindly supplied by the Rev. E. P. Stock), and she did perfectly remember her baptism, there being *seven of the family all baptized at once.* In the early part of her life she lived as domestic servant in various farmhouses in the neighbourhood, and among others with Mr. Browne, attorney-at-law, Troutbeck, great-grandfather of the present Mr. George Browne. This would be about the year 1784, when she was 20 years of age ; and it was then she said she tasted her first cup of tea, which was introduced at the christening of one of Mr. Browne's children—a fact which shows that luxury had been some time in finding its way in the then secluded vales of Westmoreland. But Peggy never kindly took to it, preferring her oatmeal porridge and meat and potatoes. *At the age of* 27 she married James Longmire, of Crawmires, Troutbeck, a remarkable yeoman, 15 years her senior. He died January 19, 1831, and had, somehow or other, managed to part with his property; consequently Peggy was obliged to depend upon her own exertions for a livelihood. This she did by acting as nurse to sick persons, an occupation which she followed up to her eightieth year; since which time she has depended upon a small amount of parochial relief and the kind help of her friends. She had only three children (sons): the eldest died about five years ago; the second and third, we believe, still live. The latter emigrated, and so at the present time we are not aware of the exact amount of her descendants; but three years since she had 10 grandchildren living in England (one of them, Thomas Longmire, the celebrated champion wrestler of England), 33 great-grandchildren, and these last will probably be much increased since that time. It is notable, by the way, that her father was 68, and her mother a little over 70, at the time of their deaths; her maternal grandmother, however, attained her 100th year; consequently she is not the first old person in the family. Since the cold weather set in last winter, she has, in a great measure, kept her bed, *but has retained all her faculties until the last, with the exception of a slight deaf-*

ness ; her memory was good until the very day of her death, and, unlike many old persons, she could remember events of recent times ; but, owing to her position, she did not take much interest in historical and political events. Peggy opened the local ball on the occasion of the marriage of the Prince of Wales in 1863. Her Majesty, on becoming acquainted with her circumstances, forwarded 3*l.* to the poor old woman in the autumn of last year. The baptismal register of Mrs. Longmire exists in Windermere parish church, and that she lived all her long life in the neighbourhood is very surprising. Her remains were interred on Wednesday last, in that part of the new burial ground which has been recently added to the old existing one of Jesus Chapel, Troutbeck, being the first interment which has taken place since its consecration. A large and respectable company attended to pay their last tribute of respect to the deceased. We understand the ceremony was performed by the Rev. W. Sewell, whose age, added to that of the deceased, amounted to 192 years. We believe it is the intention of the inhabitants to erect a suitable tablet over her remains.'

As the Atkinsons were said to be Friends, and I knew the accuracy with which the Registers of Births, &c., of the Society of Friends were kept, and had experienced, when looking into the case of Hannah Lightfoot, the readiness with which Mr. Hoyland, the obliging keeper of them at the Friends' Meeting House, Houndsditch, then assisted me, I felt assured that I might avail myself of his kindness, and very easily settle the question of Peggy's real age.

As I anticipated, I no sooner explained to Mr. Hoyland the object of my visit to him than he made a thorough search in the register of the district in which Peggy was born. It did not contain registries of the births of any children of a John and *Margaret* Atkinson; but there were entries of the births of several children of John and *Mary* Atkinson: and the following is a copy of my memoranda of these births. I have added the names of the months—Friends, as it is known, calling them only 'First Month,' &c. :—

Children of John and Mary Atkinson.

1. Sarah . . . . 1752, 5/15 (May 15).
2. George . . . . 1754, 11/19 (November 19).
3. John . . . . 1758, 10|28 (October 28).
4. Elizabeth . . . . . . 1762, 7/10 (July 10).
5. James . . . . 1764, 8/17 (August 17).
6. Thomas } twins . . . 1767, 9/18 (September 18).
7. Mary }
8. William . . . . . . 1770, 1/27 (January 27).
9. *Margaret* . . . . . 1772, 2/2 (February 2).
10. ,, [1] . . . . . 1773, 9/13 (September 13).
11. Edward [1] . . . . . 1774, 7/29 (July 29).
12. Mary . . . . . . 1777, 10/5 (October 5).

As it seemed to me by no means impossible there was some mistake as to the Christian name of the mother, I was not without the expectation of finding that the second Margaret in this list would eventually prove to be Peggy Longmire, who, even if that were so, would have been 95 when she died.

But this expectation was not realised. When I came to inquire as to the 'Atkinsons' who were baptized in

[1] Margaret and Edward said to be of Kendal Meeting, all the rest of Newby Stowes in Strickland Monthly Meeting.

1767, I found they were not seven in number, but *nine*; which I accounted for by supposing that the parents were admitted into the Church at the same time with their children. But when the Rev. F. A. Bright sent me a list of the *nine children* of John and *Mary* Atkinson (for by a curious coincidence he accidentally wrote 'Mary' instead of 'Margaret' as the name of the mother), and I found how totally different were the names of the nine children of John and Margaret from those of the seven of John and Mary Atkinson, it was clear that Peggy did not belong to the 'Friends' family. A grandson of John and Margaret has thrown a little light upon Peggy's statement that her father and mother were Quakers, by saying he had heard 'they were Quakerish disposed.'

Children of John and Mary (Margaret) Atkinson, baptized at Windermere, May 19, 1777, *were as follows*:—

Richard, in his 21st year.	*Margaret*, in her 13th year.
Hannah, in her 19th year.	Isabel, in her 11th year.
Ann, in her 16th year.	James, in his 8th year.
John, in his 15th year.	Robert, in his 3rd year.

Thomas, no age stated.

Richard, the eldest of these children, was married at Windermere to Agnes Benson, in 1783; and as this Richard witnessed the marriage of Margaret Atkinson to James Longmire on January 8, 1798 (and the similarity of the signatures to his own marriage and to the marriage of Margaret leave no doubt of his identity), Mr. Bright justly argues that Richard was her brother, and thereby establishes the identity of Peggy with the Margaret baptized in 1777.

But Peggy's own statement raises a doubt on this very point. If there is one event in a woman's life on which her recollection might be trusted, it would surely be her marriage. Now, if she were in her 13th year when baptized in 1777, she would be *in her* 34*th*, not 27th, when married in 1798. I am bound to add that this discrepancy is accounted for by the fact that her eldest son by James Longmire was born several years before her marriage ; and that she *antedated*, so to speak, her marriage, to divert attention from that fact.

I have failed in my endeavour to ascertain the age which Peggy stated herself to be to the parish authorities when age and infirmity compelled her to apply to them for assistance ; but her name appears in the list of those who received relief in the parish of Troutbeck in 1856, and she is there stated to be 90 years of age.

Mr. Browne, of Troutbeck, to whom I am indebted for this information, sends me also some extracts from the accounts of his grandfather, the writer of the certificate lately referred to, of payments made to Peggy during the period she lived in his service; and further assures me that Peggy told him, and he has no doubt correctly, that she was living with Mr. Browne when his daughter Eleanor was born, which event took place on October 9, 1784.

Such are all, certainly all the important, particulars I have been able to collect respecting this Westmoreland Centenarian. She kept her birthday on April 15, and therefore, if she were the child baptized in May, 1777, in her 13th year, she must have been born in April, 1765, and was consequently about 103 years and six weeks old when she died, May 30, 1868.

The case is not free from difficulties, which are somewhat strengthened by the accounts we have of her physical and mental condition up to the time of her death. But if my readers feel disposed to give old Peggy the benefit of the doubt, I have nothing to say against it.

MRS. PUCKLE, *said to have died in her* 106*th year.*

The 'Times' of December 13, 1872, contained the following paragraph :—

'DEATH OF A CENTENARIAN.—A correspondent writes to us :—" There has just died at High Wych, in the parish of Sawbridgeworth, Herts, on the 9th of December, Mrs. Elizabeth Puckle, *née* Elizabeth Smith, in her 106th year. She was baptized at Eastwick, Herts, near Harlow, Essex, on the 13th of September, 1767. She was bedridden, but otherwise in possession of all her faculties till within ten days of her death, being able to read a verse of a chapter in Isaiah without spectacles and without prompting. She was remarkable for her cheerful demeanour, *plump, healthy appearance*, and lively recollection of bygone facts and occurrences. She always asserted *that she remembered walking to church in pattens to be baptized*, which would seem to add at least two years to her recorded age." '

I have long been acquainted with this case, which is an extremely interesting one ; not to me, perhaps, the less interesting from the difficulties with which it is surrounded. I consider it a very doubtful one ; though

the evidence in it is sufficient, as one of my correspondents says, 'to satisfy me,' though, he candidly
adds, 'incomplete enough to admit of your indulging
your incredulity.'

My attention had been called privately to this venerable dame by some friends, just at the time when the
following interesting notice of her appeared in the
'Times' of October 18, 1871 :—

'About a mile and a half from High Wych, a hamlet
in the parish of Sawbridgeworth, there is now living an
old lady named Elizabeth Puckle, the widow of a miller
of that name. *She is even now rosy and plump, and in
good health, being able to read her Bible without the assistance of glasses.* Her birth was registered at Eastwick, in Hertfordshire, the post town of which is Harlow,
Essex, September 13, 1767, so that she is now, by her
register, 104 years old, but by computation 106. As
she, being still lively and chatty, remembers *walking to
church in her pattens, at that time worn by everybody*, she
must then have been at least two years old, so that her
neighbours give her credit for being two years older than
her register. In early life she was in service, and she
has always been remarkable for her good temper. She
is now living in a thatched cottage, with clay walls and
a floor of bricks ; her bed is near the outer door of the
cottage, and this is generally open in fine weather. *Her
memory is excellent, but her annals are short and simple,
generally relating to her neighbours and friends. It is
quite refreshing to see this old lady, so different from most
persons of extreme old age, her smile so bright, and her*

chat about the grandfathers of septuagenarians so lively.
She is supported by her children and grandchildren, and
lives in comfort. In her younger days Mrs. Puckle filled
the situation of nursemaid in the family of the grandfather of the present Mr. Rivers, of Sawbridgeworth
Nurseries, and who is 74 years of age.'

I have in this, as in the preceding quotation, italicised
some few passages to which the reader's attention is particularly desired.

The physical condition of the old lady was so inconsistent with what might be looked for in a woman who
had attained the unexampled age of 104, that I was
very glad when I was enabled, by the introduction of a
friend, to put myself in communication with the Rev.
H. Frank Johnson, the incumbent of High Wych. That
gentleman, who is firmly convinced that Mrs. Puckle was
really of the age claimed for her, has been most kind in
searching out for me all the information which I asked
for, sparing no pains to satisfy my doubts, and to arrive
at the point for which we are equally anxious—the truth
as to the real age of Mrs. Puckle.

To Mr. Johnson's kindness I am indebted for particulars from the Eastwich Registers of the entries of the
marriage of Mrs. Puckle's father and mother, John Smith
and Susannah Ricketts, which was solemnised on January
3, 1766; and of the baptism of six of their children :—

1766. April 6. *John.*

1767. Sept. 20. Elizabeth, MRS. PUCKLE.

1772. May 17. Mary (afterwards Mrs. North), died
1816.

1777. Oct. 9. William Wright, died 1859.

1779. April 6. Sarah (afterwards Mrs. Harding), died 1859.

1783. Dec. 1830. Susan (afterwards Mrs. Mead), died 1858.

Mr. Johnson added a notice of the marriage of Elizabeth Smith to Timothy Puckle, of Stapleford, which was solemnised on April 23, 1793 ; and of the birth of their daughter on November 10, 1793.

Nothing could at first sight appear more clear and satisfactory. Mrs. Puckle was known to be a daughter of John and Susannah Smith, and the sister of their several children, and here was proof of her baptism in 1767, and consequently that she was in 1871 no less than 104 years old.

But then Mrs. Puckle's physical condition—'rosy and plump,' 'in good health,' 'able to read without spectacles,' 'her memory excellent,' and 'she generally so different from persons in extreme old age'—awakened doubts in my mind, which are strengthened and not dissipated the more consideration I give to the question.

But after some time, a small piece of information with which Mr. Johnson supplemented the particulars already quoted, gave me, I think, a clue to the truth.

After enumerating the children baptized at Eastwick, Mr. Johnson went on to mention—

'Another son, Thomas, said to have been born in 1770, and to have died in 1830, is not found in the Baptismal Register.'

Now, if the reader will turn to the list of baptisms, he

will find two marked breaks in the otherwise regular intervals between the births of these little Smiths—and exceptionally long intervals, too—one being between 1767 and 1772, and the second between 1772 and 1777. Now, as it appears perfectly clear from the other entries that the Smiths duly presented the other children for baptism, there is no reason why Thomas, if born at Eastwich, should not have been baptized there in due time after his birth, or, if any cause had led to the delay, at all events when Mary was baptized in 1772 ; unless— and here, I think, is the clue to the whole mystery— Smith, the father, who was an agricultural labourer, had left Eastwich for some neighbouring parish in search of work, in which parish the boy Thomas was born and baptized.

The birth of Thomas fills up the interval between 1767 and 1772 ; and I feel strongly impressed that a thorough search in the Baptismal Registers of the parishes in the neighbourhood of Eastwich would give us evidence of the baptism of Thomas, the burial probably of the first Elizabeth, and the baptism of a second Elizabeth, who would prove to have been closely approaching a Centenarian at the time of her death, though not to have reached the unexampled age of upwards of 105 years.

I am very sceptical on the subject of the recollection of old ladies who profess to remember walking to church to be baptized, though that recollection may be strengthened by the additional fact that they ' *walked in pattens.*'

But if Mrs. Puckle did recollect that circumstance, it is clear she could not be the Elizabeth baptized in 1767, for her christening took place only 20 months after that of John. So that, if her recollection is worth anything, it proves that she was a second Elizabeth, whose age has yet to be ascertained by the discovery of her baptismal certificate.

I made an appeal in the ' Times' of January 4, 1872, to clergymen and gentlemen who take an interest in scientific truth, living in parishes near Eastwich, to devote half an hour to a search of the registers between 1772 and 1777 for the baptism of Elizabeth, daughter of John and Susannah Smith, but the appeal has been unsuccessful; so that now all I can say is, that I cannot believe Mrs. Puckle was 105 and upwards, but am at present unable to prove the negative.

APPENDICES.

———◆◇◆———
∴

APPENDIX A.

HENRY JENKINS.

No. I.

THE following is a copy of the deposition of Henry Jenkins, taken on April 15, 1667, 'at the howse of John Staireman, in Cattericke, in the county of Yorke, on the parte and behalfe of Charles Anthoney, clerke, complaynant, against Calvert · Smythson, defendant, by virtue of a Com. directed to George Wright, Joseph Chapman, John Burnett, and Richard Fawcett, gentlemen, or to any three or two of them, for the examination of witnesses between the sayd partyes.' as printed by the Rev. Canon Raine in the 'Yorkshire Archæological and Topographical Journal,' vol. i. p. 129, from the depositions in York Castle.

Jenkins was the eighth witness.

'8. Henry Jenkins, of Ellerton-upon-Swaile, in the county of Yorke, labourer, aged one hundreth fifty and seaven, or theirabouts, sworne and examined.

'1. To the first interrogatory this deponent sayeth that he knowes the partyes, complaynant and defendant, in this suite, and hath knowne them for several yeares past.

'3 and 4. To the third and fowerth interrogatory he sayeth that all the particulars mentioned in the third interrogation . . . able and due to be payed to the vicarr of Cattericke, and that the Lordshyp or manor of Kiplin is within the parish of Catteryck, and nowe in the possession of the defendant and several other tenants, and that to this deponent's knowledge all the particulars mentioned in the . . . nid interr. were payed in kinde by one Mr. Calvert, the owner of the lordship or mannor of Kiplinge, to one Mr. Thriscross, above three score yeares . . . vicar of Cattericke, and were soe payed in kinde duringe the time of his the sayd Mr. Thriscr. . . . mitance their, and after the tythes of Kiplinge were payed in kinde to one Mr. Richard fawcett . . . many yeares together as vicar of Cattericke, aforesaid, and that this deponent never knewe of· any . . . tythes, payed by and of the owners or occupyers of the lordship or manor of Kiblinge, or any other townes or hambletts within the said parish of Cattericke, but all such particulars in the third interr. . . . were ever paid in kinde to the vicar there for the time beinge.'

No. II.

The following is a reprint of Miss Savile's account of Henry Jenkins, as recorded by Dr. Tancred Robinson, in ' The Philosophical Transactions of the Royal Society,' vol. xix. pp. 266-8, No. 221, 1696 :—

' A letter, giving an account of one Henry Jenkins, a Yorkshire man, who attained the age of 169 years. Communicated by Dr. Tancred Robinson, Fellow of the College of Physitians and Royal Society, with his remarks on it.

' "Sir,—Mr. Robinson tells me you desire the relation of Henry Jenkins, which is as followeth. When I came first to live at Bolton, it was told me, there lived in that parish a man near 150 years old; that he had sworn as a witness in a cause at York to 120 years, which the judge reproving him for, he said he was butler at that time to Lord Conyers, and they told me that it was reported his name was found in some old register of the Lord Conyers' meeneal servants ; but truly it was never in my thoughts to enquire of my Lord Darcy, whether this last particular was true or no ; for I believed little of the story for a great many years ; till one day being in my sister's kitchen, Henry Jenkins coming in to beg an alms, I had a mind to examine him ; I told him he was an old man who must soon expect to give an account to God of all he did or said ; and I desired him to tell me very truly how old he was ; on which he paused a little ; and then said to the best of his remembrance he was about 162 or 163.   I asked him what kings he remembered ? he said Henry VIII.   I asked him what public thing he could longest remember ? he said Flowden Field.   I asked him whether the king was there ? he said no, he was in France and the Earl of Surry was general.   I asked him how old he might be then ? he said, I believe I might be *between* 10 *and* 12, 'for,' says he, 'I was sent to Northallerton with a horse-load of arrows, but they sent a bigger boy from thence to the army with them.' I thought by these marks I might find something in histories, and looked in an old chronicle that was in the house, and I did find that Flowden Field was 152 years before, so that if he was 10 or 11 years old, he must be 162 years or 163 as he said, when I examined him.   I found by the book that bows and arrows were then used, and that the Earl he named was then general, and that king Henry VIII. was then at Tournay, so that I don't know what to answer to the consistencies of these things, for Henry Jenkins was a poor man and could neither write nor read.   There were also four or five in the same parish that were reputed all of them to be 100 years old or within two or three years of it, and they all said he was an elderly man ever since they knew him, for he was born in another parish and before any registers were in churches as it is said ; he told me then too that he was butler to the Lord Conyers, and remembered the Abbot of Fountain's Abby very well, who used to drink a glass with his lord heartily, and that the dissolution of the monasteries he said he well remembered.   " ANN SAVILE."

'This Henry Jenkins died Dec. 8, 1670, at Ellerton-upon-Swale. The battle of Flowden Field was fought upon the 9th of Sept., in the year of our Lord 1513; Henry Jenkins was 12 years old when Flowden Field was fought, so he lived 169 years. Old Parre lived 152 years 9 months, so that Henry Jenkins outlived him by computation 16 years, and was the oldest man born upon the ruines of this post-diluvian world.

'This Henry Jenkins, in the last century of his life, was a fisherman, and used to wade in the streams ; his diet was coarse and sower ; but towards the latter end of his days, he begged up and down ; he hath sworn in Chancery and other courts to above 140 years' memory, and was often at the Assizes at York, whither he generally went a-foot, and I have heard some of the country gentlemen affirm that he frequently swum in the rivers after he was past the age of 100 years.

' 'Tis to be wished that particular enquiries were made and answered, concerning the temperament of this man's body, his manner of living, and all other circumstances which might furnish any useful instructions to those who are curious about Longævity.'

APPENDIX B.

HARRISON'S ACCOUNT OF OLD PARR.

No. I.

'Some account of Thomas de Temporibus, alias Old Tom Parr, who died November 5, 1635, extracted from a MS. chronology of Mr. Harrison, a painter, in Norfolk, now in the hands of the Rev. Francis Burton, Fellow of Pembroke Hall in Cambridge, by the Rev. John Jones, of Abbot's Ripton, in com. Hunt.

" 1. Thomas Parr died the 5 of November, 1635 (11 Car. I.). The summer before, the Earl of Arundel was at Wem, in Shropshire, and sent for the said Parr (where I saw him and spoke with him), who had then been blind nineteen years. And after two days, the said Earle sent him in a litter to the King. And

the King said to old Parr, "You have lived longer than other
men, what have you done more than other men?" He answered, "I did Penance when I was an hundred years old."
The same he told me before he went to the King.'—'Peck's
Collection of Historical Pieces.' Lond.: 1740, quarto, p. 51.

No. II.

Taylor's ' Life of Old Parr,' as here reprinted, has been carefully collated with the copy of the first edition in the British
Museum.[1]

The following extract from the Registers of the Stationers'
Company shows that the book was not entered until after Parr's
death :—

' 1635, 7 Dec.

' Henry Gosson.

' Entered for his Copy under the hand of Henry Walley,
  a Pamphlett called The old old very old Man, &c., by
  John Taylor .   .   .   .   .   . vjd.'

The book is entitled :

' THE OLD, OLD, VERY OLD MAN :

OR,

THE AGE AND LONG LIFE OF THOMAS PARR,

' *The son of John Parr, of Winnington, in the parish of Alberbury, in the*
county of Salop, or Shropshire,

' Who was born in the reign of King Edward the Fourth, [and
' is now living in the Strand,][2] being aged one hundred and
' fifty-two years and odd months ; his manner of life and conver-
' sation in so long a pilgrimage ; his marriages ; and his bringing
' up to London about the end of September last, 1635 ; where-

---

[1] With this work is bound up a broadside, entitled 'The Wonder of
this Age ; or, The Picture of a Man Living, who is 152 years old and
upward, this 12th day of November, 1635. London. Printed for Ben-
jamin Fisher, 1635.' It contains an account of Parr's habits of life. In
the centre is a Portrait of Old Parr, by C. V. Dalen, sculpt.

[2] Omitted on the title of the 1st edition.

' unto is added a postscript, shewing the many remarkable
' accidents that hapned in the life of this old man. Written
· by John Taylor. London : Printed for Henry Gosson, at his
' shop on London Bridge, neere to the Gate, 1635. 1794.

' To the High and Mighty Prince, Charles,

' By the Grace of God, King of Great Britain, France, and
Ireland, Defender of the Faith, &c.

' Of subjects, my dread liege, 'tis manifest,
You have the old'st, the greatest, and the least :
That for an old, a great, and little man,
No kingdom, sure, compare with Britain can ;
One,[1] for his extraordinary stature,
Guards well your gates, and by instinct of nature,
As he is strong, is loyal, true, and just,
Fit, and most able, for his charge and trust.
The other's small and well composed feature
Deserves the title of a pretty creature ;
And doth, (or may,) retain as good a mind
As greater men, and be as well inclin'd :
He may be great in spirit, though small in sight,
Whilst all his best of service is delight.
The old'st, your subject is ; but for my use,
I make him here, the subject of my muse :
And as his aged person gain'd the grace,
That where his Sovereign was, to be in place,
And kiss your royal hand ; I humbly crave,
His life's description may acceptance have.
And as your Majesty hath oft before
Look'd on my poems ; pray read this one more.
' Your Majesty's
' Most Humble Subject
' and Servant,
' JOHN TAYLOR.

[1] The king's gigantic porter, who once drew Jeffery, the dwarf, out of
his pocket, in a masque at court.

'The Occasion of this Old Man's being brought out
of Shropshire to London.

As it is impossible for the sun to be without light, or fire
to have no heat ; so is it undeniable that true honour is as in-
separably addicted to virtue, as the steel to the loadstone ;
and without great violence, neither the one or the other can
be sundered. Which manifestly appears, in the conveying
out of the country of this poor ancient man ; monument, I
may say, and almost miracle of nature.

' For the Right Honourable Thomas Earl of Arundel and
Surrey, Earl Marshall of England, &c. being lately in Shropshire to visit some lands and manors, which his Lordship
holds in that county ; or, for some other occasions of importance, which caused his Lordship to be there, the report of
this aged man was certified to his honour; who, hearing of so
remarkable a piece of antiquity, his Lordship was pleased to
see him ; and in his innated noble and christian piety, he took
him into his charitable tuition and protection : commanding
that a litter and two horses, for the more easy carriage of a
man so enfeebled and worn with age, to be provided for
him ' also, that a daughter-in-law of his, named Lucy, should
likewise attend him ; and have a horse for her own riding
with him ; and, to cheer up the old man, and make him
merry, there was an antique-faced-fellow, called Jack, or John
the fool, with a high and mighty no beard, that had also a
horse for his carriage. These all were to be brought out of
the country to London, by easy journies ; the charges being
allowed by his Lordship, and likewise one of his honour's
own servants, named Brian Kelley, to ride on horseback with
them, and to attend and defray all manner of reckonings
and expences ; all which was done accordingly, as followeth.

Winnington is a hamlet in the Parish of Alberbury, near a
place called the Welsh Poole, eight miles from Shrewsbury,

from whence he was carried to Wim, a town of the Earl's aforesaid; and the next day to Shefnall, a manor house of his Lordship's, where they likewise staid one night; from Shefnall they came to Wolverhampton, and the next day to Brimicham, from thence to Coventry; and although Master Kelley had much to do to keep the people off that pressed upon him in all places where he came, yet at Coventry he was most oppressed; for they came in such multitudes to see the old man, that those that defended him, were almost quite tired and spent, and the aged man in danger to have been sti..ed; and in a word, the rabble were so unruly, that Bryan was in doubt he should bring his charge no further; so greedy are the vulgar to hearken to or gaze after novelties. The trouble being over, the next day they past to Daventry, to Stony Stratford, to Redburn, and so to London, where he is well entertained and accommodated with all things, having all the aforesaid attendants at the sole charge and cost of his Lordship.

One remarkable passage of the old man's policy must not be omitted or forgotten, which is thus; his three leases of sixty-three years being expired, he took his last lease of his landlord, one Master John Porter, for his life; with which lease, he hath lived more than fifty years, as is further hereafter declared; but this old man would, for his wife's sake, renew his lease for years, which his landlord would not consent unto; wherefore old Parr, having been long blind, sitting in his chair by the fire, his wife looked out of the window, and perceived Master Edward Porter, the son of his landlord, to come towards their house, which she told her husband, saying, Husband, our young landlord is coming hither : Is he so? said old Parr; I pr'ythee, wife, lay a pin on the ground near my foot, or at my right toe; which she did; and when young Master Porter, yet forty years old, was come into the house, after salutations between them, the old man said, Wife, is not that a pin which lies at my foot? Truly husband, quoth she, it is a pin indeed; so she took up the pin, and Master Porter was half in a maze that the old man had recovered his sight again; but it was quickly found to be a witty conceit, thereby to have them to suppose him to be more lively than he was ; because he hoped ˙to have his lease renewed for his wife's sake, as aforesaid.

' He hath had two children by his first wife, a son and a daughter; the boy's name was John, and lived but ten weeks; the girl was named Joan, and she lived but three weeks. So that it appears he hath outlived the most part of the people that are living near there, thrée times over.

'THE VERY OLD MAN : OR, THE LIFE OF THOMAS PARR.

' An old man's twice a child, the proverb says, .
And many old men ne'er saw half his days,
Of whom I write ; for he at first had life,
When York and Lancaster's domestic strife
In her own blood had factious England drench'd,
Until sweet peace those civil flames had quench'd.
When as fourth Edward's reign to end drew nigh,
John Parr, a man that liv'd by husbandry,
Begot this Thomas Parr, and born was he,
The year of fourteen hundred eighty-three.
And as his father's living and his trade,
Was plough and cart, sithe, sickle, bill, and spade ;
The harrow, mattock, flail, rake, fork, and goad,
And whip, and how to load, and to unload ;
Old Tom hath shew'd himself the son of John,
And from his father's function hath not gone.

' Yet I have read of as mean pedigrees,
That have attain'd to noble dignities :
Agathocles, a potter's son ; and yet
The kingdom of Sicilia he did get.
Great Tamberlain, a Scythian shepherd was,
Yet, in his time, all princes did surpass.

First Ptolemy, the king of Ægypt's land,
A poor man's son of Alexander's band.
Dioclesian, Emperor, was a scrivener's son,
And Proba, from a gard'ner, th' empire won.
Pertinax was a bondman's son, and wan
The empire ; so did Valentinian,
Who was the offspring of a rope-maker,
And Maximinus of a mule-driver :
And if I on the truth do rightly glance,
Hugh Capet was a butcher, king of France.
By this I have digrest, I have exprest
Promotion comes not from the east or west.

‘ So much for that, now to my theme again :
This Thomas Parr hath liv'd th' expired reign
Of ten great kings and queens, th' eleventh now sways
The sceptre, blest by th' Ancient of all days.
He hath surviv'd the Edwards, fourth and fifth ;
And the third Richard, who made many a shift
To place the crown on his ambitious head ;
The seventh and eighth brave Henries both are dead;
Sixth Edward, Mary, Philip, El'sabeth,
And blest remember'd James, all these by death
Have changed life, and almost 'leven years sincè
The happy reign of Charles our gracious prince,
Tom Parr hath liv'd, as by record appears,
Nine months, one hundred fifty, and two years.
Amongst the learn'd, 'tis held in general
That every seventh year's climacterical,
And dang'rous to man's life, and that they be
Most perilous, at th' age of sixty-three,
Which is, nine climactericals ; but this man
Of whom I write, since first his life began,
Hath liv'd of climactericals such plenty,
That he hath almost out liv'd two-and-twenty.

For by records, and true certificate,
From Shropshire late, relations do relate,
That he liv'd seventeen years with John his father,
And eighteen with a master, which I gather
To be full thirty-five; his sire's decease
Left him four years possession of a lease;
Which past, Lewis Porter, gentleman, did then
For twenty-one years grant his lease again :
That lease expir'd, the son of Lewis, call'd John,
Let him the like lease ; and that time being gone,
Then Hugh, the son of John, last nam'd before,
For one-and-twenty years, sold one lease more.
And lastly, he hath held from John, Hugh's son,
A lease for's life these fifty years, out run :
And till old Thomas Parr, to earth again
Return, the last lease must his own remain.
Thus having shewn th' extension of his age,
I'll shew some actions of his pilgrimage.

' A tedious time a batchelor he tarried,
Full eighty years of age before he married :
His continence, to question I'll not call,
Man's frailty's weak, and oft doth slip and fall.
No doubt but he in fourscore years might find,   ·
In Salop's county, females fair and kind :
But what have I to do with that ? let pass ;
At the age aforesaid he first married was
To Jane, John Taylor's daughter ; and 'tis said,
That she, before he had her, was a maid.
With her he lived years three times ten and two,
And then she died, as all good wives will do.
She dead, he ten years did a widower stay,
Then once more ventur'd in the wedlock way :
And in affection to his first wife Jane,
He took another of that name again ;

With whom he now doth live, she was a widow
To one nam'd Anthony, and surnam'd Adda ;
She was, as by report it doth appear,
Of Gillsel's parish, in Montgomeryshire,
The daughter of John Lloyde, corruptly Flood,
Of ancient house, and gentle Cambrian blood.

' But hold, I had forgot, in's first wife's time,
He frailly, foully, fell into a crime,
Which richer, poorer, older men, and younger,
More base, more noble, weaker men, and stronger,
Have fallen into.
The Cytherean, or the Paphean game,
That thund'ring Jupiter did oft inflame ;
Most cruel cut-throat Mars laid by his arms,
And was a slave to love's enchanting charms ;
And many a Pagan god, and semi-god,
The common road of lustful love hath trod :
For, from the emperor to the russet clown,
All states each sex, from cottage to the crown,
Have in all ages, since the first creation,
Been foil'd, and overthrown with love's temptation :
So was old Thomas, for he chanc'd to spy
A beauty, and love entred at his eye ;
Whose powerful motion drew on sweet consent ;
Consent drew action, action drew content ;
But when the period of those joys were past,
Those sweet delights were sourly sauc'd at last.
The flesh retains what in the bone is bred,
And one colt's tooth was then in old Tom's head ;
It may be he was gull'd, as some have been,
And suff'red punishment for others sin ;
For pleasure's like a trap, a gin, or snare,
Or, like a painted harlot, seems most fair ;

But when she goes away, and takes her leave,
No ugly beast so foul a shape can have.
Fair Katherine Milton was this beauty bright,
Fair like an angel, but in weight too light ;
Whose fervent feature did inflame so far
The ardent fervour of old Thomas Parr,
That for law's satisfaction, 'twas thought meet
He should be purg'd by standing in a sheet ;
Which aged, he, one hundred and five year,
In Alberbury's parish church did wear.
Should all that so offend, such penance do,
Oh whát a price would linen rise unto ;
All would be turn'd to sheets ; our shirts and smocks,
Our table linen, very porters frocks,
Would hardly 'scape transforming ; but all's one,
He suffer'd, and his punishment is done.

' But, to proceed more serious in relation,
He is a wonder, worthy admiration,
He's in these times fill'd with iniquity,   ·
No antiquary, but antiquity ;
For his longevity's of such extent,
That he's a living mortal monument.
And as high tow'rs, that seem the sky to shoulder,
By eating time, consume away, and moulder,
Until at last in piece meal they do fall,
Till they are buried in their ruins all :
So this old man, his limbs their strength have left,
His teeth all gone but one, his sight bereft,
His sinews shrunk, his blood most chill and cold,
Small solace, imperfections manifold :
Yet still his sp'rits possess his mortal trunk,
Nor are his senses in his ruins shrunk ;
But that his hearing's quick, his stomach good,
He'll feed well, sleep well, well digest his food.

He will speak heartily, laugh, and be merry ;
Drink ale, and now and then a cup of sherry ;
Loves company, and understanding talk,
And on both sides held up, will sometimes walk ;
And though old age his face with wrinkles fill,
He hath been handsome, and is comely still ;
Well fac'd ; and though his beard not oft corrected,
Yet neat it grows, not like a beard neglected.
From head to heel, his body hath all over
A quick-set, thick-set nat'ral hairy cover,
And thus, as my dull weak invention can,
I have anatomiz'd this poor old man.
  ' Though age be incident to most transgressing,
Yet time well spent, makes age to be a blessing ;
And if our studies would but deign to look,
And seriously to ponder nature's book,
We there may read, that man, the noblest creature,
By riot and excess doth murder nature.
This man ne'er fed on dear compounded dishes,
Of metamorphos'd beasts, fruits, fowls, and fishes ;
The earth, and air, the boundless ocean
Were never rak'd nor forag'd for this man ;
Nor ever did physician to his cost,
Send purging physick through his guts in post ;
In all his lifetime he was never known,
That drinking others healths, he lost his own ;
The Dutch, the French, the Greek, and Spanish grape,
Upon his reason never made a rape ;
For riot, is for Troy an anagram ;
And riot wasted Troy with sword and flame :
And surely that which will a kingdom spill,
Hath much more pow'r one silly man to kill ;
Whilst sensuality the palate pleases,
The body's fill'd with surfeits, and diseases ;

By riot, more than war, men slaughter'd be,
From which confusion this old man is free.
He once was caught in the venereal sin,
And, being punished, did experience win ;
That careful fear his conscience so did strike,
He never would again attempt the like.
Which, to our understandings may express,
Men's days are shorten'd through lasciviousness ;
And that a competent ˉcontenting diet
Makes men live long, and sleep in quiet.
Mistake me not, I speak not to debar
Good fare of all sorts : for all creatures are
Made for man's use, and may by man be us'd,
Not by voracious gluttony abus'd.
For he that dares to scandal or deprave
Good house-keeping : oh hang up such a knave ;
Rather commend what is not to be found,
Than injure that which makes the world renown'd,
Bounty hath got a spice of lethargy,
And liberal, noble, hospitality,
Lies in consumption, almost pin'd to death,
And charity benumb'd, ne'er out of breath.
May England's few good house-keepers be blest
With endless glory, and eternal rest ;
And may their goods, lands, and their happy seed,
With heav'n's best blessings multiply and breed.
'Tis madness to build high with stone and lime
Great houses, that may seem the clouds to climb,
With spacious halls, large galleries, brave rooms,
Fit to receive a king, peers, squires and grooms ;
Amongst which rooms, the devil hath put a witch in,
And made a small tobacco-box the kitchen ;
For covetousness the mint of mischief is,
And christian bounty the highway to bliss.

To wear a farm in shoe-strings, edg'd with gold,   '
And spangled garters worth a copyhold :
A hose and doublet, which a lordship cost ;
A gaudy cloak, three manors price almost ;
A beaver, band, and feather for the head,
Priz'd at the churches tythe, the poor man's bread,
For which the wearers are fear'd, and abhor'd,
Like Jeroboam's golden calves adored.
  ' This double, treble aged man, I wot,
Knows and remembers when these things were not ;
Good wholesome labour was his exercise,
Down with the lamb, or with the lark would rise ;
In mire and toiling sweat he spent the day,
And to his team he whistled time away ;
The cock his night-clock, and till day was done,
His watch, and chief sun dial, was the sun.
He was of old Pythagoras' opinion,
That green cheese was most wholesome with an onion,
Coarse Mesclin bread ; and for his daily swigg,
Milk, butter-milk, and water, whey, and whig :
Sometimes metheglin, and by fortune happy ;
He sometimes sip'd a cup of ale most nappy,
Cyder, or perry, when he did repair
T'a Whitsun ale, wake, wedding, or a fair ;
Or when in Christmas time he was a guest
At his good landlord's house amongst the rest ;
Else he had little leisure time to waste,
Or at the ale-house, huff-cap ale to taste ;
Nor did he ever hunt a tavern fox ;
Ne'er knew a coach, tobacco, or the pox ;
His physic was good butter, which the soil
Of Salop yields, more sweet than candy oil ;
And garlick he esteem'd above the rate
Of Venice treacle, or best mithridate ;

He entertain'd no gout, no ache he felt ;
The air was good, and temp'rate where he dwelt.
Whilst mavisses,[1] and sweet tongu'd nightingales
Did chant him roundelays, and madrigals.

   ' Thus living within bounds of nature's laws,
Of his long lasting life may be some cause.
For though th' Almighty all men's days do measure,
And doth dispose of life and death at pleasure ;
Yet nature being wrong'd, man's days and date
May be abridg'd, and God may tollerate.

   ' But had the father of this Thomas Parr,
His grandfather, and his great grandfather,
Had their life's threads so long a length been spun,
They, by succession, might from sire to son
Have been unwritten chronicles, and by
Tradition shew time's mutability.
Then Parr might say he heard his father well
Say that his grandsire heard his father tell
The death of famous Edward the Confessor,
Harold, and William Conq'rour his successor ;
How his son Robert won Jerusalem,
O'ercame the Saracens, and conquer'd them ;
How Rufus reign'd, and's brother Henry next,
And how usurping Stephen this kingdom vext ;
How Maud the Empress, the first Henry's daughter,
To gain her right, fill'd England full of slaughter ;
Of second Henry's Rosamond the fair,
Of Richard Cœur de Lion ; his brave heir,
King John ; and of the foul suspicion
Of Arthur's death, John's elder brother's son ;
Of the third Henry's long reign, sixty years,
The barons war, the loss of wrangling peers ;

---

[1] Thrushes.

How Long-shanks did the Scots and French convince,
Tam'd Wales, and made his hapless son their prince.
How second Edward was Carnarvon call'd,
Beaten by Scots, and by his queen enthral'd ;
How the third Edward fifty years did reign,
And t' honour'd garter's order did ordain ;
Next how the second Richard liv'd and died ;
And how fourth Henry's faction did divide
The realm with civil, most uncivil war,
'Twixt long contending York and Lancaster ;
How the fifth Henry sway'd ; and how his son
Sixth Henry, a sad pilgrimage did run.
Then of fourth Edward, and fair mistress Shore,
King Edward's concubine, Lord Hastings' ——
Then how fifth Edward, murther'd with a trick
Of the third Richard; and then how that Dick
Was by seventh Henry slain at Bosworth field ;
How he and's son, th' eighth Henry, here did wield
The sceptre ; how sixth Edward sway'd ;
How Mary rul'd ; and how that royal maid
Elizabeth did govern, best of dames,
And phœnix-like expir'd ; and how just James,
Another phœnix, from her ashes claims
The right of Britain's sceptre as his own ;
But changing for a better, left the crown,
Where now 'tis, with King Charles ; and may it be
With him, and his most blest posterity
Till time shall end : be they on earth renown'd,
And after with eternity be crown'd.
Thus had Parr had good breeding, without reading,
He, from his sire, and grandsire's sire, proceeding,
By word of mouth might tell most famous things,
Done in the reigns of all those queens and kings.
But he in husbandry hath been brought up,
And ne'er did taste the Heliconian cup ;

He ne'er knew hist'ry, nor in mind did keep
Ought, but the price of corn, hay, kine, or
  sheep.
Day found him work, and night allow'd him rest ;
Nor did affairs of state his brain molest.
His high'st ambition was, a tree to lop,
Or, at the furthest, to a may-pole's top ;
His recreation, and his mirth's discourse,
Hath been the piper, and the hobby-horse.
And in this simple sort, he hath with pain
From childhood liv'd to be a child again.
'Tis strange, a man, that is in years so grown,
Should not be rich ; but to the world 'tis known,
That he that's borne in any land or nation,
Under a twelve-pence planet's domination,
By working of that planet's influence,
Shall never live to be worth thirteen pence.
Whereby, although his learning cannot show it,
He's rich enough to be, like me, a poet.
  ' But ere I do conclude, I will relate
Of reverend age's honourable state ;
Where shall a young man good instructions have,
But from the ancient, from experience grave ?
Roboam, son and heir to Solomon,
Rejecting ancient counsel, was undone
Almost ; for ten of the twelve tribes fell
To Jeroboam King of Israel.
And all wise princes and great potentates
Select and chuse old men, as magistrates ;
Whose wisdom, and whose reverend aspect,
Knows how and when to punish or protect.
The patriarchs' long lives before the flood,
Were given them, as 'tis rightly understood,
To store and multiply by procreations,
That people should inhabit and breed nations.

X

That th' ancients their posterities might show
The secrets deep of nature how to know,
To scale the sky with learn'd astronomy ;
And sound the ocean's deep profundity ;
But chiefly how to serve, and to obey
God, who made them out of slime and clay.
Should men live now, as long as they did then,
The earth could not sustain the breed of men ;
Each man had many wives, which bigamy
Was such increase to their posterity.
That one old man might see before he died,
That his own only offspring had supplied
And peopled kingdoms.
But now so brittle's the estate of man,
That in comparison, his life's a span.
Yet since the flood it may be proved plain,
That many did a longer life retain
Than him I write of; for Arpachshad liv'd
Four hundred thirty-eight ; Shelah surviv'd
Four hundred thirty-three years ; Eber more,
For he liv'd twice two hundred sixty-four.
Two hundred years Terah was alive,
And Abraham liv'd one hundred seventy-five.
Before Job's troubles, holy writ relates,
His sons and daughters were at marriage states ;
And after his restoring, 'tis most clear,
That he surviv'd one hundred forty year.
John Buttadeus, if report be true
Is his name that is stil'd, The Wand'ring Jew.
'Tis said, he saw our Saviour die ; and how
He was a man then, and is living now ;
Whereof relations you that will may read ;
But pardon me, 'tis no part of my creed.
Upon a German's age, tis written thus,
That one Johannes de Temporibus

Was armour bearer to brave Charlemaign,
And that unto the age he did attain,
Of years three hundred sixty-one, and then
Old John of Times return'd to earth again.
And noble Nestor, at the siege of Troy,
Had liv'd three hundred years both man and boy.
Sir Walter Rawleigh, a most learned knight,
Doth of an Irish countess, Desmond write,
Of seven score years of age, he with her spake :
The Lord Saint Albans doth more mention make,
That she was married in fourth Edward's reign,
Thrice shed her teeth, which three times came again.
The highland Scots and the wild Irish are
Long liv'd with labour hard, and temperate fare.
Amongst the barbarous Indians, some live strong
And lusty, near two hundred winters long !
So as I said before, my verse now says,
By wronging nature, men cut off their days.
Therefore as times are, he I now write on,
The age of all in Britain hath outgone ;
All those that were alive when he had birth,
Are turn'd again unto their mother earth ;
If any of them live, and do reply,
I will be sorry, and confess I lye.
For had he been a merchant, then perhaps
Storms, thunder-claps, or fear of after-claps,
Sands, rocks, or roving pirates, gusts and storms,
Had made him long ere this the food of worms.
Had he a mercer, or a silkman been,
And trusted much, in hope great gain to win,
And late and early striv'd to get or save,
His grey head long ere now had been i'th' grave.
Or had he been a judge or magistrate,
Or of great counsel in affairs of state ;

Then day's important business, and night's cares,
Had long ere this, lutter'd his hoary hairs;
But as I writ before, no cares opprest him,
Nor ever did affairs of state molest him.
Some may object, that they will not believe
His age to be so much, for none can give
Account thereof, time being past so far,
And at his birth there was no register.
The register was ninety seven years since
Giv'n by th' eighth Henry, that illustrious prince,
Th' year fifteen hundred forty wanting twaine,
And in the thirtieth year of that king's reign :
So old Parr now, was almost an old man,
Near sixty ere the register began.
I have writ as much as reason can require,
How Times did pass, how Leases did expire;
And gentlemen o' th' county did relate
T' our gracious king by their certificate
His age, and how time with grey hairs hath crown'd him ;
And so I leave him older than I found him.'

No. III.

Harvey's Autopsy of Parr, from the Sydenham Society's
edition of the *Works of William Harvey, M.D. Translated
from the Latin, with life of the Author, by Robert Willis, M.D.*
8vo, 1847, pp. 587-592.

'Thomas Parr, a poor countryman, born near Winnington, in
the county of Salop, died on the 14th of November, in the year
of grace 1635, after having lived one hundred and fifty-two
years and nine months, and survived nine princes. This poor
man, having been visited by the illustrious Earl of Arundel
when he chanced to have business in these parts (his lordship
being moved to the visit by the fame of a thing so incredible),

was brought by him from the country to London ; and, having been most kindly treated by the earl both on the journey and during a residence in his own house, was presented as a remarkable sight to his Majesty the King.

' Having made an examination of the body of this aged individual, by command of his Majesty, several of whose principal physicians were present, the following particulars were noted :

' The body was muscular, the chest hairy, and the hair on the fore arms still black ; the legs, however, were without hair, and smooth.

' The organs of generation were healthy, the penis neither retracted nor extenuated, nor the scrotum filled with any serous infiltration, as happens so commonly among the decrepid ; the testes, too, were sound and large ; so that it seemed not improbable that the common report was true, viz., that he did public penance under a conviction for incontinence, after he had passed his hundredth year ; and his wife, whom he had married as a widow in his hundred-and-twentieth year, did not deny that he had intercourse with her after the manner of other husbands with their wives, nor until about twelve years back had he ceased to embrace her frequently.

' The chest was broad and ample ; the lungs, nowise fungous, adhered, especially on the right side, by fibrous bands to the ribs. They were much loaded with blood, as we find them in cases of peripneumony, so that until the blood was squeezed out they looked rather blackish. Shortly before his death I had observed that the face was livid, and he suffered from difficult breathing and orthopnœa. This was the reason why the axillæ and chest continued to retain their heat long after his death : this and other signs that present themselves in cases of death from suffocation were observed in the body.

' We judged, indeed, that he had died suffocated, through inability to breathe, and this view was confirmed by all the physicians present, and reported to the King. When the blood was expressed, and the lungs were wiped, their substance was beheld of a white and almost milky hue.

'The heart was large, and thick, and fibrous, and contained a considerable quantity of adhering fat, both in its circumference and over its septum. The blood in the heart, of a black colour, was dilute, and scarcely coagulated ; in the right ventricle alone some small clots were discovered.

'In raising the sternum, the cartilages of the ribs were not found harder or converted into bone in any greater degree than they are in ordinary men ; on the contrary, they were soft and flexible.

'The intestines were perfectly sound, fleshy, and strong, and so was the stomach : the small intestines presented several constrictions, like rings, and were muscular. Whence it came that, by day or night, observing no rules or regular times for eating, he was ready to discuss any kind of eatable that was at hand ; his ordinary diet consisting of sub-rancid cheese, and milk in every form, coarse and hard bread, and small drink, generally sour whey. On this sorry fare, but living in his home, free from care, did this poor man attain to such length of days. He even ate something about midnight shortly before his death.

'The kidneys were bedded in fat, and in themselves sufficiently healthy ; on their anterior aspects, however, they contained several small watery abscesses or serous collections, one of which, the size of a hen's egg, containing a yellow fluid in a proper cyst, had made a rounded depression in the substance of the kidney. To this some were disposed to ascribe the suppression of urine under which the old man had laboured shortly before his death ; whilst others, and with greater show of likelihood, ascribed it to the great regurgitation of serum upon the lungs.

'There was no appearance of stone either in the kidneys or bladder.

'The mesentery was loaded with fat, and the colon, with the omentum, which was likewise fat, was attached to the liver, near the fundus of the gall-bladder; in like manner the colon was adherent from this point posteriorly with the peritoneum.

'The viscera were healthy; they only looked somewhat white externally, as they would have done had they been parboiled; internally they were (like the blood), of the colour of dark gore.

'The spleen was very small, scarcely equalling one of the kidneys in size.

'All the internal parts, in a word, appeared so healthy, that had nothing happened to interfere with the old man's habits of life, he might perhaps have escaped paying the debt due to nature for some little time longer.

'The cause of death seemed fairly referrible to a sudden change in the non-naturals, the chief mischief being connected with the change of air, which through the whole course of life had been inhaled of perfect purity,—light, cool, and mobile—whereby the præcordia and lungs were more freely ventilated and cooled; but in this great advantage, in this grand cherisher of life, this city is especially destitute; a city whose grand characteristic is an immense concourse of men and animals, and where ditches abound, and filth and offal lie scattered about, to say nothing of the smoke engendered by the general use of sulphureous coal as fuel, whereby the air is at all times rendered heavy, but much more so in the autumn than at any other season. Such an atmosphere could not have been found otherwise than insalubrious to one coming from the open, sunny, and healthy region of Salop; it must have been especially so to one already aged and infirm.

'And then for one hitherto used to live on food unvaried in kind, and very simple in its nature, to be set at a table loaded with variety of viands, and tempted not only to eat more than wont, but to partake of strong drink, it must needs fall out that the functions of all the natural organs would become deranged. Whence the stomach at length failing, and the excretions long retained, the work of concoction proceeding languidly, the liver getting loaded, the blood stagnating in the veins, the spirits frozen, the heart, the source of life, oppressed, the lungs infarcted, and made impervious to the ambient air, the general habit rendered more compact, so that it could no longer exhale or perspire—no wonder that the soul, little content with such a prison, took its flight.

'The brain was healthy, very firm and hard to the touch; hence, shortly before his death, although he had been blind for twenty years, he heard extremely well, understood all that was said to him, answered immediately to questions, and had perfect apprehension of any matter in hand; he was also accustomed to walk about, slightly supported between two persons. His memory, however, was greatly impaired, so that he scarcely recollected anything of what had happened to him when he was a young man, nothing of public incidents, or of the kings or nobles who had made a figure, or of the wars or troubles of his earlier life, or of the manners of society, or of the prices of things—in a word, of any of the ordinary incidents which men are wont to retain in their memories. He only recollected the events of the last few years. Nevertheless, he was accustomed, even in his hundred and thirtieth year, to engage lustily in every kind of agricultural labour, whereby he earned his bread, and he had even then the strength required to thrash the corn.'

No. IV.

The inscription which marked the resting-place of old Parr has lately been carefully re-engraved by order of the present Dean of Westminster, and is as follows:—

Tho: Parr of ỹ County of Sallop Borne in A°: 1483. He lived in ỹ reignes of Ten Princes viz: K. Edw. 4. K. Ed. 5. K. Rich. 3. K. Hen. 7. K. Hen. 8. K. Edw. 6. Q. Ma. Q. Eliz. K. Ja. & K. Charles Aged 152 yeares. & was Buried Here Novemb. 15. 1635.

To the courtesy of the Dean, to whom I had applied for a copy of the register of Parr's burial, I am indebted for the following note :—

'Deanery, Westminster.

'July 3rd, 1869.

'Dear Mr. Thoms,—There is no entry of Parr's burial. The early register before 1660 is very imperfect, and for the year 1635 records only two interments—Horatio Vere and Mrs. Paul.

'Yours sincerely,

'A. P. STANLEY.'

APPENDIX C.

BETTY EVANS'S TOMBSTONE.

When treating on Monumental Inscriptions as evidence of Longevity (*ante*, p. 49), I referred to the case of Betty Evans, at Pinner, in support of my view. The case was investigated by the late Mr. Dilke, and the following statement of the result appeared in the 'Athenæum' of January 3, 1857 :—

'*Longevity.*—In a review of Dr. Webster's "Statistics of Graveyards," in the "Athenæum" of the 29th of November, 1856, an allusion is made to the case of Betty Evans, who lived and died at Pinner at the age of 102, as one for which "it is probable that good evidence might be procured." As I myself lately made inquiry into this very moderate instance of Longevity, and as the result has some bearing upon the value of tombstone testimony, I will here state it for the benefit of your readers. Betty Evans's monument in Pinner Churchyard bears the following inscription :—" This stone was erected by private subscription in memory of Betty, widow of William Evans, of this parish, and daughter of William and Ann Weatherby. Born at Ruislip, 27th April, 1751. Died at Pinner, 10th of

August, 1853." The facts are here plain and precise; and when I add that the admirers of Old Betty, anticipating the reasonable test proposed by your reviewer, actually searched the parish register at Ruislip for the date of her birth before placing it upon the monument, it may naturally be concluded that we have here a sure case of 102 as a starting-point for our inquiries. Unfortunately, however, an attempt to procure corroborative testimony has only served to shake my faith even in the monument. Her daughter, Mrs. Redman, is still living at Eastcot, a village in the neighbourhood, and she informs me that when her mother died her own relatives did not know her age, although "folks" said she was more than a hundred. The father and mother of the venerable lady had been dead more than sixty years, and little, of course, could be known of them beyond a traditional account of Old Betty herself, that they lived at Ruislip, and that their name was Weatherby. Upon this information the promoters of the monument searched the Register of Baptisms, and found the evidence on which it was stated that Mrs. Redman's mother's age must be 102. The name of Weatherby, however, is very common in that part of the country; and on examining the register at Ruislip myself I found no less than three Elizabeths, the earliest of whom would probably have been considered by an enthusiastic collector of cases of Longevity as the most eligible. The dates and names are as follows :— 1740. Elizabeth, daughter of John Weatherby; 1751. Elizabeth, daughter of William and Ann Weatherby; 1766. Elizabeth, daughter of James and Elizabeth Weatherby. On the other hand, Mrs. Redman assures me that she had always understood, and frequently been told by her mother, that she was christened "Betty," and not Elizabeth. Assuming, however, that she was one of these three, why must she have been number two? The *William* Weatherby referred to was, I learn, a farmer residing at Field End. Betty's father, like her husband and sons, was a labourer—a "sheep shearer." But the grandson of William and Ann Weatherby of Field End is

still living at Eastcot, where his father lived. He told me he
knew old Betty for fifty or sixty years as an inhabitant of that
part, and that she was in no way related to him or connected
with his family. This is confirmed by old Mr. George, of Pinner, who is eighty years of age, and who knew both Betty Evans
and "Will Weatherby" for many years. Again, Mrs. Redman
had repeatedly heard her mother say that she was married at
eighteen. I could not find the register of her marriage, but her
first child, who is now living, must, according to Mrs. Redman's
statement, have been born about 1783, when, if Betty was as
old as her tombstone alleges, she was thirty-two years of age.
Thus stands the case, which appears to be inconsistent with the
dates of every one of the three Elizabeths in the register. Your
readers will probably think that the circumstances justify a
doubt, even of the simple and circumstantial account of Old
Betty engraved upon her tombstone in Pinner Churchyard.'

L'Envoy.

I have now brought to a close a volume which has extended
to a greater length than I anticipated, and which has been completed under no slight difficulty, owing to the defective state of
my eyesight—a condition very prejudicial to all literary work,
but more especially to a work of this character, in which so
much depends upon strict accuracy in dates, names, and other
minute particulars. Indeed, I doubt whether I could have persevered in it at this time but for the favourable—I fear too
favourable—opinion of its usefulness with which my distinguished friend Professor Owen has been kind enough to
encourage me.

Should my sight be mercifully restored to me, and the book
receive such encouragement as to lead to a second edition, I
shall hope to render such edition still more complete, by the
insertion of other Proved, Disproved, and Doubtful cases which
are still in progress of examination ; and by availing myself of

any suggestions for its general improvement which my critics and friends may point out.

As I have been influenced in its preparation by one object only—the advancement of Truth—so I trust I have not written one line of it in the spirit of a partisan.

INDEX.

LONDON : PRINTED BY
SPOTTISWOODE AND CO., NEW-STREET SQUARE
AND PARLIAMENT STREET

www.ingramcontent.com/pod-product-compliance
Lightning Source LLC
Chambersburg PA
CBHW021124270326
41929CB00009B/1034